Library of
Davidson College

ADVANCES IN CRYPTOLOGY

Proceedings of Crypto 82

ADVANCES IN CRYPTOLOGY
Proceedings of Crypto 82

Edited by
DAVID CHAUM
University of California
Santa Barbara, California

RONALD L. RIVEST
and
ALAN T. SHERMAN
Massachusetts Institute of Technology
Cambridge, Massachusetts

PLENUM PRESS • NEW YORK AND LONDON

Library of Congress Cataloging in Publication Data

Workshop on the Theory and Application of Cryptographic Techniques (1982: University of California, Santa Barbara)
   Advances in cryptology.

"Proceedings of a Workshop on the Theory and Application of Cryptographic Techniques, held August 23–25, 1982, at the University of California, Santa Barbara, California" — T.p. verso.
   Bibliography: p.
   Includes indexes.
   1. Computers — Access control — Congresses. 2. Cryptography — Congresses. I. Chaum, David. II. Rivest, Ronald L. III. Sherman, Alan T. IV. Title.
QA76.9.A25W67 1982   001.64   83-9492
ISBN 0-306-41366-3

Proceedings of a Workshop on the Theory and Application of Cryptographic Techniques, held August 23–25, 1982, at the University of California, Santa Barbara, California

©1983 Plenum Press, New York
A Division of Plenum Publishing Corporation
233 Spring Street, New York, N.Y. 10013

All rights reserved

No part of this book may be reproduced, stored in a retrieval system, or transmitted in any form or by any means, electronic, mechanical, photocopying, microfilming, recording, or otherwise, without written permission from the Publisher

Printed in the United States of America

CRYPTO 82

A Workshop on the Theory and Application of Cryptographic Techniques

held at the University of California, Santa Barbara

August 23–25, 1982

with the cooperation of
the IEEE Communications Society,
the IEEE Information Theory Group,
and the Department of Computer Science
at the University of California, Santa Barbara

Organizers

David Chaum (UCSB), general chairman
Leonard M. Adleman (USC), program committee
Thomas A. Berson (SYTEK), Hatfield conference coordinator
Dorothy Denning (Purdue), program committee
Whitfield Diffie (BNR), program committee
Paul Eggert (UCSB), treasurer
Allen Gersho (UCSB), program committee
John Gordon (Hatfield Polytechnic), organizing committee
David Kahn (Cryptologia), organizing committee
Richard Kemmerer (UCSB), local arrangements chairman
Stephen Kent (BBN), program committee
John Kowalchuk (MITRE), registration
Ronald L. Rivest (MIT), program committee chairman
Alan T. Sherman (MIT), program committee assistant chairman
Stephen Weinstein (AMEX), organizing committee

Preface

In the opening sentence of their seminal 1976 paper, Diffie and Hellman proclaimed: "We stand today on the brink of a revolution in cryptography."[1] Six years later, we find ourselves in the midst of this revolution, surrounded by an explosion of developments in cryptology.

Cryptology is the art of making and breaking codes and ciphers. More generally, cryptology provides techniques for transmitting information in a private, authenticated, and tamper-proof manner. Cryptology was once the exclusive domain of mathematicians, governments, and military forces. But as computer and communications technologies advance, and as we move toward an electronically interconnected society, more and more people now depend on computer mail, electronic business transactions, and computer data banks. Cryptology has become a vital concern of numerous businesses and individuals. Fortunately, the availability of small, fast, and inexpensive computers has made encryption feasible and economical for many applications.

Organized in response to the growing interest in cryptology, CRYPTO 81 was the first major open conference ever devoted to technical cryptologic research.[2] Its successor, CRYPTO 82, was the largest conference of its kind. Held August 23–25, 1982, CRYPTO 82 attracted over 100 participants, including many leading researchers from all over the world. CRYPTO 82 took place at the University of California at Santa Barbara and was held with the cooperation of the IEEE Communications Society, the IEEE Information Theory Group, and the Department of Computer Science at U. C. Santa Barbara.[3] Compiled as the official record of

---

[1] Whitfield Diffie and Martin E. Hellman, "New Directions in Cryptography," *IEEE Transactions on Information Theory*, IT–22 (November 1976), 644.

[2] Held August 24–26, 1981, CRYPTO 81 took place at the University of California at Santa Barbara. It was sponsored by the IEEE Data and Computer Communications Committees and was supported in part by the National Science Foundation. The CRYPTO 81 proceedings are available as a technical report: Allen Gersho, ed., "Advances in Cryptology: A Report on CRYPTO 81," ECE Report no. 82-04, Department of Electrical and Computer Engineering, U. C. Santa Barbara, Santa Barbara, California 93106.

[3] Additional details about the conference can be found in: David Kahn, "The CRYPTO 82 Conference, Santa Barbara: A Report on a Conference," *Cryptologia*, 7 (January 1983), 1–5.

CRYPTO 82, *Advances in Cryptology: Proceedings of CRYPTO 82* helps to document the current explosion in cryptologic research.

This volume contains 34 papers that were presented at CRYPTO 82, as well as a paper by Donald W. Davies from CRYPTO 81 that did not appear in the CRYPTO 81 proceedings. Most of these papers appear here in print for the first time. As a unique record of the current state of cryptologic research, *Advances in Cryptology: Proceedings of CRYPTO 82* is an invaluable source of information for anyone intrigued by the recent developments in cryptology. *Advances in Cryptology* is also well suited for use as a supplementary textbook in a course in cryptology.

Reflecting the structure of the conference, the proceedings are arranged in six sections. The first five sections contain the main papers of the conference, organized roughly according to the following themes: algorithms and theory, modes of operation, protocols and transaction security, applications, and cryptanalysis. The sixth section contains abstracts describing results presented at the informal "Rump Session."

Each paper in the five main sections was selected by the program committee from brief abstracts submitted in response to a call for papers. The final papers were not formally refereed, and the authors retain full responsibility for the contents of their papers. Several of the papers are preliminary reports of continuing research.

Section I, "Algorithms and Theory," focuses on specific cryptographic algorithms used to encipher messages and on theoretical foundations for the design of secure algorithms. Many of the papers in this section have a number-theoretic flavor.

Section II, "Modes of Operation," explores two major topics: the security of the Data Encryption Standard (DES) and the use of randomization to increase the security of cryptographic algorithms. For example, papers by Donald W. Davies and Robert J. Jueneman investigate the security of DES when used in output feedback mode. The underlying theme of this section is that the security provided by a cryptographic algorithm is determined in part by the way the algorithm is used.

Section III, "Protocols and Transaction Security," studies how protocols can be used to conduct various business transactions electronically. In particular, protocols are discussed for signing checks, making untraceable payments, and enabling two mutually suspicious parties to sign a contract simultaneously. Methods for proving the correctness of such protocols are also examined in detail.

Section IV, "Applications," treats the key management aspects of a number of cryptographic applications, such as protecting personal data cards, controlling access to local networks, and implementing an electronic notary public. This section also includes a paper by Charles Bennett *et al.* suggesting that quantum mechanics, rather than computational complexity, can form the foundation for certain cryptographic schemes.

Section V, "Cryptanalysis," investigates weaknesses of knapsack ciphers. In what is perhaps the most significant unclassified cryptologic paper of the year, Adi Shamir explains how to break the basic Merkle-Hellman knapsack public-key cryptosystem. Gustavus J. Simmons *et al.* and Leonard M. Adleman describe related discoveries. During the conference, Adleman's presentation was particularly notable for his use of an *Apple II* personal computer to solve an instance of the Graham-Shamir knapsack cipher. Jeff Legarias and Donald W. Davies also presented papers, but these papers were not received in time to be included in the proceedings. Davies's talk, which concluded the session, was a fascinating overview of techniques used by the Allies during WWII to break the *Enigma* cryptograph.

Section VI, "Rump Session," covers a potpourri of cryptologic topics including DES, multi-party protocols, pseudo-random number generators, threshold schemes, randomized stream ciphers, and the RSA nMOS chip. The papers in this section summarize brief impromptu talks given at an informal evening session of the conference.

Unfortunately, a few papers presented at CRYPTO 82 could not be included in this book. A list of these papers appears immediately following the table of contents.

During his opening remarks, David Chaum proposed the formation of an international organization that would further research in cryptology by coordinating and organizing meetings in the area, and by informing its members of relevant events, publications, and work. During CRYPTO 82, attendees nominated a planning committee. The members were Henry J. Beker, Ernest F. Brickell, David Chaum, Whitfield Diffie, Robert R. Jueneman, David Kahn, and Stephen Kent. At the end of the conference, the planning committee held its first meeting, at which it adopted the working title: "The International Association for Cryptologic Research." The committee also laid plans for a meeting in Europe during March 1983 and for a CRYPTO 83 during August 1983.

The editors would like to thank all of the authors, organizers, and other people who made these proceedings possible. We are grateful to Leonard M. Adleman, Dorothy Denning, Whitfield Diffie, and Stephen Kent, who served as session chairmen and members of the program committee. We also thank Allen Gersho for his help on the program committee. Several other people made essential contributions to the conference: John Kowalchuk handled registration; Richard Kemmerer took care of local arrangements; Paul Eggert served as treasurer; and Thomas A. Berson coordinated CRYPTO 82 with the Hatfield conference. We would also like to express our appreciation to John Gordon, David Kahn, and Stephen Weinstein for helping organize the conference. Finally, our thanks go to Inna Sverbilov of MIT and L. S. Marchand of Plenum Press for their patient and cheerful assistance in preparing this book.

Santa Barbara, California   D.C.
Cambridge, Massachusetts   R.L.R.
January 1983   A.T.S.

Contents

Session I: Algorithms and Theory
Ronald L. Rivest, chairperson

Fast Computation of Discrete Logarithms in $GF(q)$ 3
    Martin E. Hellman and Justin M. Reyneri

Some Remarks on the Herlestam-Johannesson Algorithm
for Computing Logarithms over $GF(2^p)$ 15
    Ernest F. Brickell and J. H. Moore

A Public-Key Cryptosystem Based on the Matrix Cover
NP-Complete Problem 21
    Ravi Janardan and K. B. Lakshmanan

Infinite Structures in Information Theory 39
    G. R. Blakley and Laif Swanson

A Fast Modular Multiplication Algorithm with Applications
to Two Key Cryptography 51
    Ernest F. Brickell

Comparison of Two Pseudo-Random Number Generators 61
    Lenore Blum, Manuel Blum, and Michael Shub

On Computationally Secure Authentication Tags Requiring
Short Secret Shared Keys 79
    Gilles Brassard

Session II: Modes of Operation
Dorothy Denning, chairperson

Some Regular Properties of the 'Data Encryption Standard'
Algorithm *(Presented at CRYPTO 81)* 89
   Donald W. Davies

The Average Cycle Size of the Key Stream in Output
Feedback Encipherment *(Abstract)* 97
   Donald W. Davies and G. I. P. Parkin

Analysis of Certain Aspects of Output Feedback Mode 99
   Robert R. Jueneman

Drainage and the DES 129
   Martin E. Hellman and Justin M. Reyneri

Security of a Keystream Cipher with Secret Initial Value *(Abstract)* 133
   Robert S. Winternitz

Using Data Uncertainty to Increase the Crypto-Complexity of
Simple Private Key Enciphering Schemes 139
   G. M. Avis and S. E. Tavares

Randomized Encryption Techniques 145
   Ronald L. Rivest and Alan T. Sherman

Session III: Protocols and Transaction Security
Leonard M. Adleman, chairperson

On the Security of Multi-Party Protocols
in Distributed Systems 167
   Danny Dolev and Avi Wigderson

On the Security of Ping-Pong Protocols *(Extended Abstract)* 177
   Danny Dolev, Shimon Even, and Richard M. Karp

The Use of Public-Key Cryptography for Signing Checks 187
   Luc Longpré

Blind Signatures for Untraceable Payments 199
    David Chaum

A Randomized Protocol for Signing Contracts *(Extended Abstract)* 205
    Shimon Even, Oded Goldreich, and Abraham Lempel

On Signatures and Authentication 211
    Shafi Goldwasser, Silvio Micali, and Andy Yao

Session IV: Applications
Stephen Kent, chairperson

Cryptographic Protection of Personal Data Cards 219
    Christian Mueller-Schloer and Neal R. Wagner

Non-Public Key Distribution 231
    Rolf Blom

Cryptographic Solution to a Multilevel Security Problem 237
    Selim G. Akl and Peter D. Taylor

Local Network Cryptosystem Architecture: Access Control 251
    Thomas A. Berson

Implementing an Electronic Notary Public 259
    Leonard M. Adleman

Quantum Cryptography, or Unforgeable Subway Tokens 267
    Charles H. Bennett, Gilles Brassard, Seth Breidbart, and Stephen Wiesner

Session V: Special Session on Cryptanalysis
Whitfield Diffie, chairperson

A Polynomial Time Algorithm for Breaking the
Basic Merkle-Hellman Cryptosystem *(Extended Abstract)* 279
    Adi Shamir

A Preliminary Report on the Cryptanalysis of
Merkle-Hellman Knapsack Cryptosystems 289
    Ernest F. Brickell, J. A. Davis, and Gustavus J. Simmons

On Breaking the Iterated Merkle-Hellman Public-Key Cryptosystem 303
    Leonard M. Adleman

Rump Session: Impromptu Talks by Conference Attendees
Alan Sherman, chairperson

Long Key Variants of DES 311
    Thomas A. Berson

On the Security of Multi-Party Ping-Pong Protocols *(Abstract)* 315
    Shimon Even and Oded Goldreich

Inferring a Sequence Produced by a Linear Congruence *(Abstract)* 317
    Joan B. Plumstead

Key Reconstruction *(Abstract)* 321
    Michael Merritt

Nondeterministic Cryptography 323
    Carl Nicolai

A Short Report on the RSA Chip 327
    Ronald L. Rivest

Author Index 329

Subject Index 331

The following papers were presented at CRYPTO 82, but were not received in time to be included in this publication:

Session III: Protocols and Transaction Security

Encryption and Protocols
    Stephen Kent (Bolt, Beranek, and Newman, Inc.)

Session V: Special Session on Cryptanalysis

The Strongest Knapsack-Based Cryptosystem?
    Adi Shamir (Weizmann Institute, Israel)

On Simultaneous Diophantine Approximation Problems
    Jeff Lagarias (Bell Laboratories)

The Bombe at Bletchley Park
    Donald W. Davies (National Physical Laboratory, England)

Rump Session: Impromptu Talks by Conference Attendees

A Short Report on the Hatfield Conference
    John Gordon (Hatfield Polytechnic, England)

What Would You Do if You Got a Secrecy Order?
    Pete Olwell (International Phasor Telecom)

Session I: Algorithms and Theory

FAST COMPUTATION OF DISCRETE LOGARITHMS IN $GF(q)$

Martin E. Hellman and Justin M. Reyneri

Information Systems Laboratory
Stanford University
Stanford, California 94305

Abstract

The Merkle-Adleman algorithm computes discrete logarithms in $GF(q)$, the finite field with q elements, in subexponential time, when q is a prime number p. This paper shows that similar asymptotic behavior can be obtained for the logarithm problem when $q = p^m$, in the case that m grows with p fixed. A method of partial precomputation, applicable to either problem, is also presented. The precomputation is particularly useful when many logarithms need to be computed for fixed values of p and m.

Introduction

If α is a primitive element in $GF(q)$, the Galois field with q elements, then the pair of inverse functions

$$Y = \alpha^X$$

and

$$X = \log_\alpha(Y)$$

are well defined over $GF(q)$ for $0 \leq X \leq (q-2)$ and Y ranging over the nonzero field elements. The first is the discrete exponential function while the second is the discrete logarithm, or index function.

This pair of inverse functions has long been of interest to number theorists, and has recently gained practical importance due to applications in cryptography. [1,2,3] Their application to cryptography stems from the apparent one-way nature of the discrete exponential function -- it is easy to compute but appears hard to invert. The difficulty of computing discrete logarithms is of crucial concern to the

---

This work was supported by the National Security Agency under contract number MDA904-81-C-0414 and by the National Science Foundation under grant number ECS-79-16161.

security of several proposed cryptosystems and key distribution systems, both conventional and public key.

Merkle [4,5] and Adleman [6] independently developed the fastest known algorithm for computing discrete logarithms for the case $q=p$, a prime. Adleman showed it to have a running time which grows subexponentially in the length of the problem l, where $l=m\ln(p)$. The bound he obtained was roughly

$$\exp\{\sqrt{k l \ln(l)}\}.$$

This grows slower than any exponential in l, but faster than any polynomial. (For this algorithm, $m=1$ and we believe $k\approx 4$.)

This paper discusses extensions of the algorithm to cases where $q=p^m$, for certain values of p and m. Our method achieves subexponential asymptotic behavior when m grows with p fixed. The method is of particular interest because $GF(2^m)$ has been suggested to have certain advantages for the public key distribution system suggested by Diffie and Hellman. [2,7]

To our knowledge, there is no subexponential solution of the discrete logarithm problem for the case when m and p are arbitrary, including, for example, logarithms in $GF(p^2)$..

The Merkle-Adleman Algorithm

We begin with a brief discussion of Merkle's original method for $GF(p)$, best motivated by the following three observations. (Recall that elements of $GF(p)$ can be represented as integers from 0 to $p-1$, with addition and multiplication done modulo p. In what follows, it often is convenient to delete the modularity notation, leaving it tacit. Because $\alpha^{p-1}=\alpha^0=1$, arithmetic in the exponent is done modulo $p-1$ and the sums below are performed in that system.)

1) If the discrete logarithm problem is "easy" for some small but non-negligible fraction ϵ of the elements Y_i in $GF(p)$, it can not be much harder for the others. Consider the problem for a particular element, Y_0. With probability ϵ, $\log_\alpha Y_0$ is easy to compute, and we are done. If not, choose r_1 uniformly from $[0,p-2]$ and let $R_1=Y_0\alpha^{r_1}$. If $\log_\alpha R_1$ is easily computed, then $\log_\alpha Y_0=\log_\alpha R_1-r_1 \mod (p-1)$. If not choose $r_2,r_3,...$ independently until an easy R_i is found and compute Y_0 by subtraction. Each new R_i has independent probability ϵ of success so the expected effort is $1/\epsilon$ times the effort for each trial value R_i.

2) If the logarithms of the first M primes, $p_1,...,p_M$, are known and if an integer R_i factors completely into $p_1,...,p_M$, then $\log_\alpha R_i$ can be computed by addition:

$$R_i = \prod_{j=1}^{M} p_j^{R_{ij}} \mod p \quad iff \quad \log_\alpha R_i = \sum_{j=1}^{M} R_{ij}\log_\alpha p_j \mod p-1 \ .$$

Adleman refers to integers which factor completely into small primes as "smooth" numbers. Checking whether a number is smooth is not difficult, so we see that $\log_\alpha R_i$ is easily computed if R_i is smooth and the logarithms of the small primes are known.

3) Finally, Merkle proposed computing the logarithms of the small primes in the following way. Choose integers r uniformly and independently from $[0, p-2]$ and exponentiate until a smooth R_1 is found with corresponding exponents $\{R_{ij}\}$, $1 \leq j \leq M$. This yields the relationship

$$\alpha^{r_1} = R_1 = \prod_{j=1}^{M} p_j^{R_{1j}} \mod p$$

or

$$r_1 = \log_\alpha R_1 = \sum_{j=1}^{M} R_{1j} \log_\alpha p_j \mod p-1.$$

Continue the process for $R_2, R_3, \ldots$ to obtain M independent equations, then solve the resulting system for the unknowns $\log_\alpha p_j$.

The notation used above is used in the sequel, so we take a moment to explain it more fully. Singly-subscripted lower case letters are used as integer exponents of primitive elements, and the result of the exponentiation is denoted by the corresponding upper case letter, as in $R_1 = \alpha^{r_1}$. We may also write $R_1 = Y_0 \alpha^{r_1}$, where Y_0 is the field element whose logarithm we wish to find -- the meaning should be clear from context. In either case, when R_1 is smooth, we use doubly-subscripted upper case letters for the exponents of the small primes in the factorization of R_1, and a singly-subscripted bold face letter for the associated vector. For example, we might have $\alpha^{r_1} = R_1 = \prod_{j=1}^{M} p_j^{R_{1j}}$, together with $\mathbf{R}_1 = (R_{11}, \cdots, R_{1M})$.

The number of small primes M is a parameter of the algorithm which is chosen to optimize performance. We will not discuss the optimization in detail in this paper, but will briefly mention some of the tradeoffs involved. If M is made too small, the probability ϵ that a randomly chosen integer R factors completely into the first M primes becomes unacceptably small, and the time to find such R's dominates the running time of the algorithm. If M is too large, checking a random R for smoothness dominates the computation. Generating an invertible matrix also takes longer, for we need at least M smooth R's and their associated vectors before we can invert the matrix. In addition, as M grows, the dimension of the vectors increases, and so do memory requirements and the number of operations needed for matrix inversion.

Notice that Merkle's algorithm naturally divides into two parts. The first is precomputation of the logarithms of the small primes, which need only be done once for a particular prime p. Once the precomputation is complete, logarithms can be computed using only the randomization procedure described in observation 1) above. In practice, this would probably be the best approach to take if many logarithms need to be computed for a single value of p. Unfortunately, we cannot prove subexponential behavior for this method. The problem lies in the requirement for an invertible system of equations to compute the $\log_\alpha p_i$. It is not difficult to predict the rate at which smooth R_i are generated by the randomization process, but it is not clear how many such R_i are needed to get an independent set of equations. It is conceivable that the vectors $\mathbf{R}_i = (R_{i1},...,R_{iM})$ resulting from each R_i would tend to be degenerate, making the time to find a spanning set prohibitive.

Adleman's algorithm [6] overcomes this difficulty by eliminating the need for an invertible set of equations. This allows him to provide a rigorous theoretical treatment, but the proof that he outlines does not show that the precomputation, which dominates the postcomputation, can be re-used. For our algorithm we provide a different proof which shows that we can have both subexponential behavior and the flexibility of precomputation. The new proof applies equally well to $GF(p)$ as $p \to \infty$ and $GF(p^m)$ as $m \to \infty$ with p fixed.

From the above brief discussion, it is apparent that the density of smooth integers is crucial to the performance of the Merkle-Adleman algorithm. Since elements of $GF(p^m)$ are most conveniently represented as polynomials, we will extend the definition of smooth to polynomials. The density of smooth quantities will remain crucial to the performance of the extended algorithm.

Extension to $GF(p^m)$

Elements of $GF(p^m)$ can be represented by polynomials $g(D)$ of degree $m-1$ or less with integer coefficients in the range $[0,p-1]$. Addition of field elements is done by adding the polynomials term by term, and reducing the coefficients mod p. Multiplication begins with the usual polynomial multiplication followed by reduction of coefficients mod p. For convenience in what follows, we will refer to the process described so far, with its single modular reduction, as "normal" multiplication. To effect "modular" multiplication, terms of degree m or greater are then reduced modulo an irreducible polynomial $f(D)$ of degree m. For example, consider multiplying $D+1$, an element of $GF(2^2)$, by itself. We take $f(D) = D^2 + D + 1 = 0$ for the the irreducible polynomial. Since $1 = -1 \ mod \ 2$, $D^2 + D + 1 = 0$ is equivalent to $D^2 = D + 1$. Thus normal multiplication proceeds as

$$(D+1)(D+1) = D^2 + 2D + 1 = D^2 + 1.$$

Modular multiplication then reduces this result modulo $f(D)$ to obtain

$$D^2 + 1 = D + 1 + 1 = D.$$

The idea behind our extension is simple: we need to find, among these polynomials, analogs to the small prime integers used in the basic algorithm. We can then proceed much as before. We must of course show that these analogs share with prime integers the properties needed to guarantee subexponential behavior. As mentioned earlier, the analogy is only sufficiently close for $m >> p$.

Our analogs to the small prime integers are monic irreducible polynomials of low degree. In "normal" polynomial multiplication the irreducibles are much like primes: Polynomials factor uniquely into monic irreducibles and a single scalar term, and the length (degree) of the product is the sum of the lengths of the factors. It is also easy to check whether a particular polynomial factors completely into a small set of irreducibles and to determine the factorization. (Because we must have this last property, we are forced to operate in normal multiplication when checking for smoothness. If a small prime irreducible happens to be a primitive element of the field, then all field elements "factor" completely when modular multiplication is used, but determining the "factorization" is exactly the discrete logarithm problem.) Finally, for $m\gg p$, we will see that the density of irreducibles is of the same form as that of the prime integers, so that the density of smooth polynomials is of the same form as that of smooth integers, where we apply "smooth" to polynomials according to the

Definition: A polynomial is smooth w.r.t k if it factors completely into a scalar and any number of monic irreducibles of degree k or less.

This definition suggests the following minor modification of our previous notation. If $\beta = \alpha^{r_1}$ is a smooth field element, we will write

$$\beta = \alpha^{r_1} = c_1 R_1, \tag{1}$$

with c_1 a scalar and R_1 a *monic* polynomial which factors into a product of small (degree $\leq k$) monic irreducibles. In this case, we write $\mathbf{R}_1$ for the vector of exponents of irreducibles in the factorization of R_1. For simplicity, we assume a table giving the logarithms of the scalars is available. There are p scalars in $GF(p^m)$, so the use of a table does not affect our asymptotic result when p is fixed, but, as we shall see below, we can get by with the computation of the logarithm of just a single scalar. If p is large, one may therefore wish to use the original Merkle-Adleman algorithm to eliminate the table of logarithms of scalars.

Like the original algorithm, our method can be broken into two parts. In the precomputation, we choose integers r at random until we find an r_1 such that α^{r_1} is smooth. We then compute $c_1 R_1 = \alpha^{r_1}$ and the corresponding vector $\mathbf{R}_1$. We continue in this way to generate a set of vectors $\{\mathbf{R}_i\}$, $1 \leq i \leq I$, which is a "virtual spanning set" with high probability, where:

Definition: A set of vectors, $\{\mathbf{R}_i\}$, $1 \leq i \leq I$, generated by a random procedure is a virtual spanning set if additional vectors generated by the same randomization procedure lie in the span of $\{\mathbf{R}_i\}$ with probability greater than 1/2.

The number of vectors I is a random variable whose value is determined by the outcome of the randomized precomputation.

In the second part of the algorithm, we compute the desired logarithm $\log_\alpha Y_0$ by choosing integers s_i at random until we find an s_0 such that

$$Y_0 \alpha^{s_0} = c_0 S_0 \tag{2}$$

is smooth and that the corresponding vector $\mathbf{S}_0$ is dependent upon $\{\mathbf{R}_i\}$. Because $\{\mathbf{R}_i\}$ is a virtual spanning set, the dependence is expected once we have produced a few smooth elements s_0. Once the dependence is found, we have

$$\mathbf{S}_0 = \sum_{i=1}^{I} a_i \mathbf{R}_i \mod p^m - 1,$$

from which, using (1), we obtain

$$\log_\alpha S_0 = \sum_{i=1}^{I} a_i \log_\alpha R_i = \sum_{i=1}^{I} a_i (r_i - \log_\alpha c_i) \mod p^m - 1.$$

From (2) we see that $\log_\alpha S_0 = \log_\alpha Y_0 + s_0 - \log_\alpha c_0$ so that

$$\log_\alpha Y_0 = -s_0 + \sum_{i=1}^{I} a_i r_i + \log_\alpha c_0 - \sum_{i=1}^{I} a_i \log_\alpha c_i \mod p^m - 1. \tag{3}$$

The values for s_0, r_i and a_i were obtained in the above computation, and the logarithms of the scalars c_0 and c_i are available in a small table. Thus (3) is sufficient to calculate $\log_\alpha Y_0$. Note that (3) can be rewritten

$$\log_\alpha Y_0 = -s + \sum_{i=1}^{I} a_i r_i + \log_\alpha (c_0 \prod_{i=1}^{I} c_i^{-a_i}) \mod p^m - 1.$$

In this form, it is clear that we only need to compute the logarithm of a single scalar to compute $\log_\alpha Y_0$, as claimed above.

To describe the precomputation in detail, we introduce the following notation. Let $k \ll m$ be the degree of the largest of the small irreducibles; its value will be specified shortly. Define $P(k)$ to be the number of irreducible polynomials of degree less than k. $P(k)$ plays the role of M in our earlier description of Merkle's Algorithm. To simplify notation, let $u = \lfloor (m-1)/k \rfloor$. Then, as shown in Appendix A, $\epsilon = \binom{P(k)+u}{u}/p^m$ lower bounds the probability that a random Y in $GF(p^m)$ is smooth w.r.t. k. Finally, define $b_1 = 4P(k)$, and $b_2 = 2b_1/\epsilon$. The first is the number of smooth elements we wish to generate, the second the number of random trials we are willing to perform to try to get them. The precomputation proceeds as follows.

As shown in Appendix B, set

$$k = \left\lfloor \frac{1}{\ln p} \sqrt{\frac{1}{3} l \ln(2m)} \right\rfloor. \tag{4}$$

to optimize the bound for the running time of the algorithm. Begin with $I=0$. By attempting at most b_2 exponentiations of random r, try to produce b_1 smooth field elements. As each smooth element is produced, compute $\alpha^r = cR$ and the vector **R**. If **R** is independent of $\{\mathbf{R}_i\}$, $1 \leq i \leq I$, add **R** to $\{\mathbf{R}_i\}$ and increase I by 1. Stop if $\{\mathbf{R}_i\}$ is a spanning set. We will refer to the value of I at the end of the precomputation as $I_{\max}$.

Note that we could theoretically stop as soon as $\{\mathbf{R}_i\}$ is a virtual spanning set. We know of no practical way to determine when this occurs, so we have chosen b_1 large enough to essentially guarantee that the desired set is produced.

We need to prove the following two theorems regarding the precomputation:

Theorem 1: With k chosen as in (4), and for any positive δ, precomputation of $\{\mathbf{R}_i\}$, $1 \leq i \leq I_{\max}$ can be done in

$$O(\exp\{(1+\delta)\sqrt{12l\ln(2m)}\})$$

operations comparable to arithmetic operations on elements of $GF(p^m)$.

Theorem 2: The set $\{\mathbf{R}_i\}$, $1 \leq i \leq I_{\max}$ produced by the precomputation fails to be a virtual spanning set with probability approaching zero as l, the length of the problem, approaches infinity.

The proof of Theorem 1 is very similar to the proof that appears in Dixon [8] showing that factoring can be done in subexponential time. In Dixon's proof, the number of primes less than a particular value was of importance, and the Prime Number Theorem comes into play. Instead, we are interested in the density of irreducibles, so the proof of Theorem 1 relies on

Lemma 1: For $k \geq 1$, $\frac{p^k}{2k} \leq P(k) \leq p^k$.

Proof of Lemma 1: To prove the first inequality, we write

$$P(k) = \sum_{i=0}^{k} I(i) \geq I(k),$$

where $I(i)$ is the number of monic irreducibles of degree i. Berlekamp [9] shows that $I(k) \geq \frac{p^k}{k}(1-p^{1-k/2})$, from which we obtain $I(k) \geq \frac{p^k}{k}(\frac{1}{2})$ for $k \geq 4$. Exact evaluation of $I(k)$ for $k=1,2,3$ completes the proof of the first inequality: Berlekamp shows $I(1)=p$, $I(2)=\frac{1}{2}(p^2-p)$, and $I(3)=\frac{1}{3}(p^3-p)$. The second inequality simply bounds the number of monic irreducibles by the total number of polynomials of degree $\leq k$.

Lemma 1 shows that number of irreducible monic polynomials in the first p^k polynomials is at least $O(p^k/k) = O(p^k/\ln(p^k))$. It is thus similar to the Prime Number Theorem, which states that the number of primes less than an integer n is $O(n/\ln(n))$. Using this result, we can derive a lower bound on the probability of smoothness for polynomials which is of the same form as that obtained by Dixon for smoothness of integers. Having done so, we follow his analysis to prove Theorem 1. The reader is referred to Appendix B for details.

Proof of Theorem 2: To upper bound the probability of failure in the precomputation, we will assume that failure occurs if either 1), we fail to produce b_1 smooth elements in b_2 random exponentiations, or 2), the set $\{\mathbf{R}_i\}$ is not a virtual spanning set even though it was produced from b_1 vectors. Let P_1 and P_2, respectively, be the probabilities associated with these events. We obtain an upper bound for overall failure by summing the bounds for P_1 and P_2. The individual bounds are obtained by applying the Chebyshev inequality to upper bound the probability that fewer than N_1 successes are obtained in N_2 Bernoulli trials with probability of success δ,

$$P(\sum_{i=1}^{N_2} x_i \leq N_1) < \frac{\delta(1-\delta)}{N_2(N_1/N_2)^2} \tag{5}$$

To obtain the bound for P_1, note that each random exponentiation in the precomputation produces a smooth element with probability greater than or equal to ϵ. The number of trials is b_2, and the desired number of smooth elements is b_1. Substituting $\delta=\epsilon$, $N_1=b_1$, and $N_2=b_2$ into (5), we find

$$P_1 \leq \frac{\epsilon(1-\epsilon)}{b_2(b_1/b_2)^2} < \frac{\epsilon}{b_2(b_1/b_2)^2} = \frac{1}{2P(k)}.$$

To bound P_2, we can consider a successful trial to be one where a vector increases the span of $\{\mathbf{R}_i\}$. The probability of a successful trial is not constant, but varies with the particular elements of $\{\mathbf{R}_i\}$. However, if $\{\mathbf{R}_i\}$ fails to be a virtually spanning set, we know that the probability of success at each trial was at least $1/2$, by definition of virtually spanning. We also know that there must have been fewer than $P(k)$ successes, for $P(k)$ is the dimension of the vectors, and $P(k)$ or more successes would mean that $\{\mathbf{R}_i\}$ was a spanning set. Substituting $\delta=1/2$, $N_1=P(k)$, and $N_2=b_1=4P(k)$ into (5) as before, we have

$$P_2 \leq \frac{1}{P(k)}.$$

Using the lower bound of Lemma 1 for $P(k)$, substituting (4) for k and simplifying, we can sum the above results to show that the probability of failure in the precomputation is less than a constant times

$$\exp\{-0.5\sqrt{l\ln(2m)}\}.$$

This completes the proof of Theorem 2.

Discussion

We have shown that the Merkle-Adleman method can be extended to $GF(p^m)$ for fixed p, while retaining subexponential behavior for the running time of the algorithm. Although the proofs we have given do not do so, they can be modified to include the case where p grows more slowly than $\ln(m)$. Other cases, for example $GF(p^2)$ as p grows, remain open. The difficulty lies in the fact that small irreducibles cannot be combined in a wide variety of ways to produce field elements; in $GF(p^2)$, we can only multiply some number of scalars times a single linear term before the resulting polynomials exceed the degree of the field. In this case, even if we assume that the logarithms of the scalars in the field are known, we can do little better than table-lookup to compute the logarithms of the first-degree terms.

The method we have presented is applicable to the public-key distribution system based on logarithms in $GF(2^{127})$ proposed by the Mitre Corporation. [7] The bound we obtained, $\exp\{(1+\delta)\sqrt{k\ln(2m)}\}$, with $k=12$, does not indicate that the algorithm is a threat, for it implies a running time of about 10^{19} years on a one-microsecond/operation machine. (In order to obtain these rough bounds, we let $\delta=0$.) With somewhat more complicated proofs, and making some reasonable assumptions about the distribution of irreducible polynomials, we believe a similar bound with $k=4$ can be obtained. This would suggest that 500,000 years are needed to break the Mitre system using the algorithm described in this paper. Use of million-fold parallelism would reduce the time to six months, and foreseeable advances in technology might reduce that time to a week in five to ten years. The security of the Mitre system is therefore marginal from a conservative point of view.

To compare the security of the RSA public-key distribution system [3] (when used as a key distribution system) with that of discrete-logarithm based systems [5], we could continue to use the estimate $k=4$. The RSA system relies for its security on the difficulty of factoring, and the best of the known algorithm for factoring is subexponential with k estimated to be 1. [11] Since the present algorithms for factoring are more efficient than those for the discrete logarithm, we might conclude that logarithm-based systems are more secure. Prudence, however, dictates that we take a lesson from the history of algorithms for factoring. While factoring has been studied much longer than the discrete logarithm problem, subexponential algorithms for factoring are relatively new, and these have improved steadily in recent years. It would seem wise to anticipate the same for the discrete-logarithm problem, and therefore to compare the two key-distribution systems assuming that $k=1$ for both. Doing so leaves us with no reason to prefer key-distribution systems based on logarithms to the RSA system. In fact, the RSA system is probably preferable, since it can also provide digital signatures and direct message protection as opposed to merely key distribution.

APPENDIX A

Define S to be the set of elements in $GF(p^m)$ which are smooth w.r.t. k, and let $u = \lfloor (m-1)/k \rfloor$. Then we have

Lemma 2: The probability that an element chosen uniformly at random from $GF(p^m)$ is an element of S is at least $\binom{P(k)+u}{u}/p^m$.

Proof of Lemma 2: Since $GF(p^m)$ contains p^m elements, we need to show that S contains at least $\binom{P(k)+u}{u}$ elements. We can produce a smooth element by multiplying together any u or fewer monic irreducible polynomials of degree not exceeding k. The result is a smooth element because it has degree at most $uk = \lfloor (m-1)/k \rfloor k \leq m-1$. We can think of a product of fewer than u factors as a product of exactly u factors by allowing 1 as a repeated factor, and therefore add the element 1 to the set of small monic irreducibles. Thus, there are at least as many smooth elements in $GF(p^m)$ as there are ways to choose u objects from a set of $P(k)+1$ objects, with replacement but without regard to order. This last number is given by $\binom{P(k)+u}{u}$ as shown, for example, in [10]. This completes the Proof of Lemma 2.

APPENDIX B

Proof of Theorem 1: The precomputation of $\{R_i\}$, $1 \leq i \leq I_{\max}$, consists of two basic procedures. In the first, we choose integers r_i randomly, calculate $Y_i = \alpha^{r_i}$ and check if Y_i is smooth. This is done at most b_2 times. The check for smoothness of the resulting Y_i can be done by trial division by all $P(k)$ small irreducibles, plus at most m additional divisions due to repeated factors. We can neglect the time required to compute $Y_i = \alpha^{r_i}$, since it can be done in time which is logarithmic in r_i. Thus, $O(b_2[P(k)+m])$ operations comparable to arithmetic operations on field elements suffice for the exponentiations and checks for smoothness.

The second basic procedure in precomputing $\{R_i\}$ checks to see if the set $\{R_i\}$ produced so far is a spanning set, and is done at most b_1 times. (Recall that there is a vector R_i associated with each smooth Y_i, and that at most b_1 smooth elements are produced in the precomputation.) The check for a spanning set can be done by a modified version of Gaussian elimination [Poh thesis] in time $O([P(k)]^3)$, since the vectors involved are of length $P(k)$. Thus, the total time spent on the precomputation is

$$O(b_2[P(k)+m]) + O(b_1[P(k)]^3) \tag{1}$$

We now upper bound each of the above terms separately, starting with

$$O(b_2[P(k)+m]) \tag{2}$$

Substituting for b_2, ignoring constant factors and noting from Lemma 1 and our choice of k that $P(k) \gg m$, (2) can be bounded by

$$O([P(k)]^2/\epsilon) = O([P(k)]^2 p^m \frac{u!P(k)!}{[P(k)+u]!}) \tag{3}$$

By applying Stirling's formula and reducing the denominator we obtain the following bounds for (3):

$$O([P(k)]^2 p^m \frac{u^u [P(k)]^{P(k)}}{[P(k)+u]^{P(k)+u}}) < O([P(k)]^2 p^m \frac{u^u}{[P(k)]^u}). \tag{4}$$

Depending on whether $P(k)$ and u appear in the numerator or denominator of (4), we use the upper or lower bounds of Lemma 1 for $P(k)$ and

$$(m/k) - 1 \leq u \leq m/k \tag{5}$$

for u to obtain the bound for (4),

$$O(\frac{p^{2k} p^m (m/k)^{m/k}}{(p^k/2k)^{(m/k)-1}}) \leq O(p^{3k}(2m)^{m/k}), \tag{6}$$

which can be rewritten as

$$O(\exp\{3k\ln p + \frac{m}{k}\ln(2m)\}). \tag{7}$$

Equation (7) is an upper bound for (1) for any positive k, and is optimized when $k = \frac{1}{\ln p}\sqrt{(1/3)l\ln(2m)}$. (Recall that $l = m\ln p$ is the length of the problem.) Since we proved Lemma 1 only for integer values of k, we choose $k = \lceil \frac{1}{\ln p}\sqrt{(1/3)l\ln(2m)} \rceil$ to obtain the following bounds.

For any $\delta > 0$, and for k sufficiently large, the following inequalities hold:

$$\frac{1}{(1+\delta)\ln p}\sqrt{(1/3)l\ln(2m)} < k \leq \frac{1}{\ln p}\sqrt{(1/3)l\ln(2m)}. \tag{8}$$

Using (8) to bound (7) results in

$$O(\exp\{\sqrt{3l\ln(2m)} + \frac{(1+\delta)m\ln(2m)\ln p}{\sqrt{(1/3)l\ln(2m)}}\}). \tag{9}$$

Substituting $l = m\ln p$ into the above and simplifying, we obtain the final bound for (2)

$$O(\exp\{(1+\delta)\sqrt{12l\ln(2m)}\}). \tag{10}$$

The second term in (1), $O(b_1[P(k)]^3) = O([P(k)]^4)$, is treated similarly. Bounding $P(k)$ by p^k and using (8) for k, we get

$$O(\exp\{4\sqrt{(1/3)l\ln(2m)}\}). \tag{11}$$

Comparing (10) and (11), we see that the former dominates, indicating that the time to find the smooth elements dominates the time to solve the resulting linear equations. We also conclude that (10) is an upper bound for the number of operations required in the precomputation, as was to be shown.

REFERENCES

[1] S. Pohlig and M. Hellman, "An improved algorithm for computing logarithms over GF(p) and its cryptographic significance," *IEEE Trans. on Inform. Theory*, vol. IT-24, pp. 106-110, Jan. 1978.

[2] W. Diffie and M. E. Hellman, "New directions in cryptography," *IEEE Trans. on Inform. Theory*, vol. IT-22, pp. 644-654, Nov. 1976.

[3] R. L. Rivest, A. Shamir and L. Adleman, "A method for obtaining digital signatures and public-key cryptosystems," *Communications of the ACM*, vol. 21, no. 2, February 1978

[4] R. Merkle, *Secrecy, Authentication, and Public Key Systems*, Ph.D. dissertation, Department of Electrical Engineering, Stanford University, June 1979.

[5] S. C. Pohlig, *Algebraic and Combinatoric Aspects of Cryptography*, Ph.D. Dissertation, Department of Electrical Engineering, Stanford University, Oct. 1977

[6] L. Adleman, "A subexponetial algorithm for the discrete logarithm with applications to cryptography," *Proceedings of the 20th Annual Symposium on Foundations of Computer Science*, Oct. 29-31, 1979

[7] S. Berkovits, J. Kowalchuk, and B. Schanning, "Implementing public-key scheme," *IEEE Commun. Mag.*, vol. 17, pp 2-3, May 1979

[8] J. D. Dixon, "Asymptotically fast factorization of integers," *Math. Comp.*, vol. 36, no. 153, Jan. 1981

[9] E.R. Berlekamp, *Algebraic Coding Theory*, New York: McGraw Hill, 1978

[10] W. Feller, *An Introduction to Probability Theory and Its Applications*, 3rd ed. New York: Wiley, 1968.

[11] R. Schroeppel, Private communication

SOME REMARKS ON THE HERLESTAM-JOHANNESSON ALGORITHM

FOR COMPUTING LOGARITHMS OVER GF(2^p)[†]

E. F. Brickell and J. H. Moore

Sandia National Laboratories
Albuquerque, New Mexico 87185

At the 1981 IEEE Symposium on Information Theory, T. Herlestam and R. Johannesson presented a heuristic method for computing logarithms over GF(2^p). They reported computing logarithms over GF(2^{31}) with surprisingly few iterations and claimed that the running time of their algorithm was polynomial in p. If this were true, the algorithm could be used to cryptanalyze the Pohlig-Hellman cryptosystem, currently in use by Mitre Corporation for key distribution. The Mitre system operates in GF(2^{127}). However, the algorithm was not implemented for GF(2^p) for p > 31 because it would require multiple precision arithmetic. Consequently attempts to evaluate the possible threat to the Pohlig-Hellman cryptosystem have centered on modeling the algorithm so that some predictions could be made analytically about the number of iterations required to find logarithms over GF(2^p) for p > 31.

The basic operation of the algorithm can be explained quite briefly. Let t be a primitive element in GF(2^p). Then α can be expressed uniquely in the form $\alpha = \sum_{i=0}^{n-1} a_i t^i$ where $a_i = 0$ or 1. The Hamming weight of α, HWT(α), is defined to be the number of a_i's that are 1 when α is expressed in the above form. Let HJ(α) = $\{t^{-2^r}\alpha^{2^s} | 0 \leq r, s \leq p-1\}$ and MINHJ(α) = min$\{$HWT(β)$| \beta \in$ HJ(α)$\}$. In the Herlestam-Johannesson algorithm, a table is constructed consisting of the logarithms of all elements $\alpha \in$ GF(2^p) with HWT(α) $\leq$ h, where h is a fixed predetermined value. Then the logarithm of an arbitrary element $\alpha \in$ GF(2^p) is found by the following iterative procedure. Let $\alpha_0 = \alpha$ and let α_1 be an element in HJ(α) such that

---

[†] This work performed at Sandia National Laboratories supported by the U. S. Department of Energy under contract number DE-AC04-76DP00789.

$HWT(\alpha_i) = MINHJ(\alpha_{i-1})$. The procedure stops when $HWT(\alpha_k) \leq h$. Then the logarithm of α can be easily computed.

The arithmetic in $GF(2^p)$ is implemented by choosing a primitive polynomial $p(t)$ of degree p over $GF(2)$ and forming the ring $GF(2)[t]/p(t)GF(2)[t]$ which is isomorphic to $GF(2^p)$. The elements of $GF(2^p)$ can then be represented as polynomials over $GF(2)$ of degree less than p with componentwise addition modulo 2 and polynomial multiplication modulo $p(t)$.

Since there is no apparent relationship between $HWT(\alpha)$ and $MINHJ(\alpha)$, previous models have assumed that they are independent. If this assumption is valid, then the algorithm would operate no better than random switching. However, using this model, after precomputation of the logarithms of Hamming weight 2, finding the logarithm of an element in $GF(2^{31})$ should require about 4500 iterations. However, Herlestam and Johannesson reported an average of 250 iterations with a maximum of 1147 in this setting. This discrepancy has been unexplained.

If $HWT(\alpha)$ and $MINHJ(\alpha)$ are independent, then for each j

$$\text{prob}[MINHJ(\alpha) = j | HWT(\alpha) = i] = \text{prob}[MINHJ(\alpha) = j]$$

should hold for all i. These probabilities can be calculated directly for $j = 1$. Let

$$HJINV(\alpha) = \{\beta | \alpha \in HJ(\beta)\} \quad .$$

By observing that

$$HJINV(\alpha) = \{t^{2^r} \alpha^{2^s} | 0 \leq r,s \leq p-1\} \quad ,$$

the computation of $HJINV(\alpha)$ is simplified. Thus, the set of all α so that $MINHJ(\alpha) = 1$ can be found by calculating $HJINV(t^k)$ for all k. From this the probability that $MINHJ(\alpha) = 1$ given that $HWT(\alpha) = i$ can be found for all i. Implementing $GF(2^{31})$ by using the primitive polynomial $t^{31} + t^3 + 1$, as was done by Herlestam and Johannesson, these probabilities were calculated and the results are shown in Table 1.

This certainly does not support the assumption that $MINHJ(\alpha)$ and $HWT(\alpha)$ are independent. Since the representation of an element, and thus its Hamming weight, is dependent upon the primitive polynomial used to implement the field operations, these calculations were repeated for various primitive polynomials. A sample of these results are shown in Table 2. The polynomial used is given in octal code.

Table 1

Polynomial is $t^{31} + t^3 + 1$

I	PROB(MINHJ(X) = 1 | HWT(X) = I)
2	.46451612
3	.06785317
4	.01945018
5	.00191865
6	.00092356
7	.00017835
8	.00008531
9	.00002495
10	.00001616
11	.00000749
12	.00000611
13	.00000361
14	.00000366
15	.00000284
16	.00000260
17	.00000217
18	.00000205
19	.00000185
20	.00000212
21	.00000211
22	.00000168
23	.00000266
24	.00000380
25	.00000679
26	.00002942
27	.00009534
28	.00066740
29	.00215053
30	0.00000000
31	0.00000000

These probabilities are essentially consistent with the assumption of independence. Define the Hamming weight of a polynomial over GF(2) to be the number of terms with a nonzero coefficient. For all polynomials tested of Hamming weight 3, the assumption of independence was not supported by these calculated probabilities but did appear valid for the 8 polynomials tested with Hamming weight between 7 and 21.

The efficiency of the Herlestam-Johannesson algorithm seems to depend upon the primitive polynomial used to implement the arithmetic. This, at first, appeared to provide a way to thwart a cryptanalytic attack using this algorithm, since changing from one primitive polynomial to another requires, in general, the solution

Table 2

Polynomial is 33201110211

I	PROB(MINHJ(X) = 1 | HWT(X) = I)
2	0.00000000
3	0.00000000
4	0.00000000
5	0.00000000
6	.00000407
7	..00000228
8	.00000494
9	.00000496
10	.00000530
11	.00000537
12	.00000507
13	.00000543
14	.00000511
15	.00000481
16	.00000489
17	.00000551
18	.00000511
19	.00000509
20	.00000472
21	.00000541
22	.00000541
23	.00000520
24	.00000266
25	.00000272
26	.00001177
27	0.00000000
28	0.00000000
29	0.00000000
30	0.00000000
31	0.00000000

of a system of multilinear equations. However, J. Sachs at Mitre Corporation, has noted that in many cases, including p = 127, this can be reduced to a system of linear equations [5]. Thus, if a polynomial is found for which the Herlestam-Johannesson algorithm works efficiently, it could remain a threat to the Pohlig-Hellman scheme.

However, the polynomial used to implement $GF(2^p)$ does affect the model used to analyze the algorithm. The results reported here indicate that a new model is needed to predict the number of iterations required for computing logarithms over larger fields. Mitre has been investigating the possibility of using a Markov process.

The underlying assumption there is that the probability that MINHJ(α) = j is dependent only on the HWT(α), which is unfortunately also somewhat suspect.

REFERENCES

1. S. Berkovits, J. Kowalchuk and B. Schanning, "Implementing a Public Key Scheme," IEEE Communications Magazine, 17, May 1979, pp. 2-3.

2. W. Diffie and M. Hellman, "New Directions in Cryptography," IEEE Trans. Inform. Theory, IT-22 (1976), pp. 644-654.

3. T. Herlestam and R. Johannesson, "On Computing Logarithms over $GF(2^p)$," BIT 21 (1981), pp. 326-334.

4. S. Pohlig and M. Hellman, "An Improved Algorithm for Computing Logarithms over GF(p) and its Cryptographic Significance," IEEE Trans. Inform. Theory, IT-24 (1978), pp. 106-110.

5. J. Sachs, private communication.

A PUBLIC-KEY CRYPTOSYSTEM BASED ON THE MATRIX COVER NP-COMPLETE PROBLEM

Ravi Janardan and K.B. Lakshmanan

Purdue University and Indian Institute of Technology, Madras

ABSTRACT

The design and implementation of a public-key cryptosystem based on the matrix cover NP-complete problem is described. Section 1 explains the problem, provides the necessary background to understand the implementation and serves to establish the terminology used. The implementation is described in detail in section 2. Section 3 contains further comments on the system and also examines its signature capability. The system borrows quite a few ideas from the Merkle-Hellman scheme [3] and a very brief comparison between the two systems appears in section 4.

1. Description Of The Problem, The Associated Background And Terminology

1.1. The Matrix Cover Problem

Given an $n*n$ matrix $A=(a_{ij})$ with nonnegative integer entries and an integer K, is there a function $f : \{1,2,...,n\} \rightarrow \{-1,1\}$ such that

$$\sum_{1 \le i,j \le n} a_{ij} f(i) f(j) \le K \ ?$$

The problem is NP-complete in the strong sense and remains so even if A is positive semidefinite [1, page 282].

For our purposes we consider the following version of the above problem :

Given an $n*n$ matrix $A=(a_{ij})$ with nonnegative integer entries and an integer K, find a function $f : \{1,2,...,n\} \to \{-1,1\}$ such that

$$\sum_{1 \leq i,j \leq n} a_{ij} f(i) f(j) = K ,$$

provided such an f exists.

This problem is at least as hard as the corresponding decision problem (with '$\leq$' replaced by '$\geq$').

Consider the special case where the matrix A is symmetric, all its entries are nonnegative and are positive powers of 2 (not necessarily consecutive powers) and the above-diagonal entries are all distinct.

Define some mapping $f : \{1,2,...,n\} \to \{-1,1\}$. Let $K = \sum_{1 \leq i,j \leq n} a_{ij} f(i) f(j)$. Since $f(i)f(j) = +1$ or -1, K is simply the sum of the a_{ij}'s with suitable signs. Note that the diagonal entries always contribute a positive sum to K. Thus, let

$$K' = K - \sum_{1 \leq i \leq n} a_{ii} f(i) f(i) = \sum_{\substack{1 \leq i,j \leq n \\ i \neq j}} a_{ij} f(i) f(j) .$$

Also, since A is a symmetric matrix, the elements a_{ij} and a_{ji}, $1 \leq i,j \leq n$, $i \neq j$, contribute equally, both in magnitude and in sign, to K'. Let

$$A_P = \{ a_{ij} \mid f(i) = f(j) , i < j \}$$

and

$$A_N = \{ a_{ij} \mid f(i) \neq f(j) , i < j \} ,$$

i.e., we are considering elements of A above the diagonal. Elements in A_P contribute positively to K' while elements in A_N contribute negatively. A_P and A_N partition the set of the above-diagonal elements of A and $|A_P| + |A_N| = (n^2 - n)/2$. So,

$$K' = 2 \sum_{a_{ij} \varepsilon A_P} a_{ij} - 2 \sum_{a_{ij} \varepsilon A_N} a_{ij} ,$$

or

$$d = K'/2 = \sum_{a_{ij} \varepsilon A_P} a_{ij} - \sum_{a_{ij} \varepsilon A_N} a_{ij} .$$

For want of a better name, we call d the Characteristic Difference (CD) of A under the mapping f.

1.2. Determining f

Our aim is to determine f for a given A and Characteristic Difference d. In general, the problem is difficult. However, as shown below, for our special choice of A it is easy.

Before proceeding further we state two lemmas. The proofs appear in Appendix A.

LEMMA 1 : Let A be symmetric matrix in which all elements are nonnegative and are positive powers of 2. Also, let the above-diagonal elements be distinct. Then, the Characteristic Difference d is always nonzero.

LEMMA 2 : If A is chosen as in lemma 1, then for a given Characteristic Difference d, the partition (A_P, A_N) is unique.

Given d, the first step in determining the mapping f is to decide which of the a_{ij}'s are in A_P and which in A_N. Lemma 2 above assures us that, given a Characteristic Difference d, we can determine the partition without ambiguity provided the matrix A satisfies the conditions set forth in that lemma.

We examine below how this partition could be obtained. We have,

$$d = \sum_{a_{ij} \varepsilon A_P} a_{ij} - \sum_{a_{ij} \varepsilon A_N} a_{ij} .$$

If d is positive then the largest element in $A_P \cup A_N$ is in A_P. If d is negative then the largest element in $A_P \cup A_N$ is in A_N. This follows from the property of sums of powers of 2. The largest element in $A_P \cup A_N$ is greater than the sum of *all* the remaining elements. The case $d = 0$ never arises (lemma 1).

For concreteness, assume that d is positive. So, the largest element a_L in $A_P \cup A_N$ is in A_P. We now have,

$$\sum_{a_{ij} \varepsilon A_P - \{a_L\}} a_{ij} - \sum_{a_{ij} \varepsilon A_N} a_{ij} = d - a_L .$$

(If d were negative we would have $a_L \varepsilon A_N$ and $\sum_{a_{ij} \varepsilon A_P} a_{ij} - \sum_{a_{ij} \varepsilon A_N - \{a_L\}} a_{ij} = d + a_L$).

We next find the second largest element in $A_P \cup A_N$ and decide (depending on whether $d - a_L$ is positive or negative) if it is in A_P or in A_N and so on. More precisely, we have :

Algorithm P :
begin
    $A_P \leftarrow \varphi$; $A_N \leftarrow \varphi$ { φ denotes the empty set }
    Let AD denote the set of the above-diagonal elements of A .
    while $|AD| \neq 0$ **do**
        begin
            Select the largest element a_L from AD.
            $AD \leftarrow AD - \{a_L\}$
            if $d > 0$ **then**
                begin
                    $A_P \leftarrow A_P \cup \{a_L\}$
                    $d \leftarrow d - a_L$
                end
            else
                begin
                    $A_N \leftarrow A_N \cup \{a_L\}$
                    $d \leftarrow d + a_L$
                end
        end
end

Algorithm P yields the (unique) partition (A_P , A_N).

For each $a_{ij} \varepsilon A_P$, $f(i)$ and $f(j)$ are either both equal to +1 or both equal to -1. For each $a_{ij} \varepsilon A_N$, either $f(i) = +1$ and $f(j) = -1$ or vice versa. We have at our disposal $|A_P| + |A_N| = (n^2 - n) / 2$ elements to help us determine the mapping f. Algorithm F below requires only $(n-1)$ of these elements to obtain f. The elements used are a_{12}, a_{13}, ..., a_{1n}.

Algorithm F :
```
begin
    f(i) ← 1
    for j ← 2 step 1 until n do
        begin
            if a_{1j} is in A_P then f(j) ← f(1)
                              else f(j) ← − f(1)
        end
end
```

Note that the values of $f(j)$, $2 \leq j \leq n$, depend on the value of $f(1)$. We could equally well have chosen $f(1)$ to be -1. So, for a given CD we get two mappings f_1 and f_2 both of which are 'correct' in that

$$\sum_{\substack{1 \leq i,j \leq n \\ i<j}} a_{ij} f_1(i) f_1(j) = \sum_{\substack{1 \leq i,j \leq n \\ i<j}} a_{ij} f_2(i) f_2(j) = d.$$

In fact, $f_1(j) = -f_2(j)$, $1 \leq j \leq n$. We defer further discussion of this to section 2.4.

It is fairly obvious that for a given CD, f_1 and f_2 are the only mappings possible. This is so because in algorithm F, once the value of $f(1)$ is fixed (at +1 or -1), the values of $f(j)$, $2 \leq j \leq n$, are determined solely by the membership of a_{1j} in A_P or in A_N. For a given CD there is only one partition (A_P, A_N), (lemma 2).

2. Implementation Of The Public-key Cryptosystem

A public-key cryptosystem based on the matrix cover problem could be designed as described below.

2.1. Generation Of The Encryption and Decryption Keys

At the time of joining the network , each user creates his encryption and decryption keys as follows :

1 . He constructs an $n*n$ secret matrix $A = (a_{ij})$ having the following properties :

(i) A is symmetric.

(ii) all elements of A are nonnegative and are positive powers of 2.

(iii) the above-diagonal elements of A are all distinct.

2 . Next, he chooses two positive integers ω and m, $\omega < m$, such that

(i) $\gcd(\omega, m) = 1$.

(ii) $m > 2 \sum_{\substack{1 \leq i,j \leq n \\ i<j}} a_{ij}$.

3 . He then computes an ω^{-1} such that $\omega\omega^{-1} \equiv 1 \pmod{m}$. The method suggested in [2, exercise 15, page 315] may be used. Since $\gcd(\omega, m) = 1$, the existence of such an ω^{-1} in the ring of integers modulo m is guaranteed. The above-diagonal a_{ij}'s, m and ω^{-1} constitute the user's secret decryption key.

4. He computes $b_{ij}' = a_{ij}\omega \bmod m$, $1 \le i, j \le n$, $i < j$. Now, since the a_{ij}'s are all powers of 2, it is possible that some of the b_{ij}''s bear a simple relation to one another. To hide this, random multiples q_{ij} of m are added to each b_{ij}' to obtain b_{ij}.[*] Thus,

$$b_{ij} = b_{ij}' + q_{ij}m, \quad 1 \le i, j \le n, i < j, q_{ij} \ge 0.$$

These b_{ij}'s constitute the above-diagonal elements of a matrix B. They form the user's publicly known encryption key. Instead of adding random multiples of m, one could iterate the $\omega - m$ transformation a few times by starting with the secret matrix $A = (a_{ij})$ and using different $\omega - m$ pairs for each iteration to arrive at the final public matrix.

2.2. The Encryption Procedure

Suppose user S wishes to send a message M to user R. Assume that M is represented as a binary string $m(1) m(2) \ldots m(n)$ of length n. S does the following:

Step E1:

He computes values $f(1), f(2), \ldots, f(n)$ using the program segment below.

begin
    for $i \leftarrow 1$ **step** 1 **until** n **do**
        begin
            if $m(i) = 1$ **then** $f(i) \leftarrow 1$
                    else $f(i) \leftarrow -1$
        end
end

Step E2:

He retrieves R's public encryption key and computes

$$C = \sum_{\substack{1 \le i,j \le n \\ i < j}} b_{ij} f(i) f(j).$$

C is the ciphertext for message M. S transmits C to R. (Actually, the ciphertext is the representation of C under some encoding scheme. However, for convenience, we will not make that distinction here and will refer to C itself as the ciphertext).

2.3. The Decryption Procedure

R recovers M from C as follows: (In what follows, rem(x, y) denotes the remainder when x is divided by y and $|z|$ represents the absolute value of z. Note that rem(x, y) is negative whenever x is).

Step D1:

R executes the following program segment:

---

[*] It was recently pointed out to the first author by Adi Shamir that adding random multiples of the modulus does not really camoflauge the secret matrix. However, it has not been possible to incorporate this change in time for the proceedings.

```
begin
     d ← rem($C\omega^{-1}$, m)
     if |d| > $\sum_{\substack{1 \leq i,j \leq n \\ i<j}} a_{ij}$ then if d > 0 then d ← d − m
                                                       else d ← d + m
end
```

Step D2 :

He uses algorithm P on his secret above-diagonal a_{ij}'s and d above to determine (A_P, A_N).

Step D3 :

He uses algorithm F to determine the values of $f(1), f(2), \ldots, f(n)$.

Step D4 :

He reconstructs M using the program segment below.
```
begin
     for i ← 1 step 1 until n do
          begin
               if f(i) = 1 then m(i) ← 1
                          else m(i) ← 0
          end
end
```

2.4. Overcoming The Message Ambiguity Problem

As already mentioned in section 1.2, algorithm F determines two mappings f_1 and f_2, both of which are correct in the sense that

$$\sum_{\substack{1 \leq i,j \leq n \\ i<j}} a_{ij} f_1(i) f_1(j) = \sum_{\substack{1 \leq i,j \leq n \\ i<j}} a_{ij} f_2(i) f_2(j) = d.$$

This means that at the end of step D4, two messages (binary strings) M_1 and M_2 are obtained. M_2 is the complement of M_1. R is faced with the problem of deciding which of M_1 and M_2 is the message transmitted by S. (The proof that one of M_1 and M_2 is the transmitted message M appears in Appendix B).

This ambiguity can be resolved in the following manner : The problem arises because of the freedom to *choose* the value of $f(1)$ in algorithm F. A network convention is established that $f(1)$ is *always* 1. This implies that the first bit in *every* message is *always* 1. Correspondingly, S encodes the information he wishes to convey to R as a binary string of length $(n-1)$. This is then 'made up' to a message M of length n by prefixing it with a '1'. The first bit serves to 'hold down' the rest of the bits when using algorithm F at step D3. At the end of step D4, R simply discards the first bit in M and obtains the information S wanted to convey. As a result of the above convention, step E1 of the encryption procedure in section 2.2 is modified slightly. Assume that the information S wishes to convey is represented as a binary string $m(2)m(3) \ldots m(n)$ of length $(n-1)$.

Step E1 (modified) :
S computes $f(1), f(2), \ldots, f(n)$ using the program segment below.
begin
$\quad f(1) \leftarrow 1$
$\quad$ **for** $i \leftarrow 2$ **step** 1 **until** n **do**
$\quad\quad$ **begin**
$\quad\quad\quad$ **if** $m(i) = 1$ **then** $f(i) \leftarrow 1$
$\quad\quad\quad\quad$ **else** $f(i) \leftarrow -1$
$\quad\quad$ **end**
end

2.5. Examples

We illustrate the working of such a public-key cryptosystem by means of two small examples.

Example 1 :

Encryption And Decryption Key Generation :

Here we assume that $n = 4$. User R chooses his (secret) matrix A to be

$$\begin{bmatrix} 2^3 & 2^1 & 2^6 \\ & 2^7 & 2^5 \\ & & 2^4 \end{bmatrix} = \begin{bmatrix} 8 & 2 & 64 \\ & 128 & 32 \\ & & 16 \end{bmatrix}.$$

Only the above-diagonal elements are shown since these are the ones of interest. Actually, only these need be retained by each user. For this case we have

$$\sum_{\substack{1 \leq i,j \leq 4 \\ i<j}} a_{ij} = 8 + 2 + 64 + 128 + 32 + 16 = 250.$$

R chooses m as 582 and ω as 7. Note that $m > 2 \sum_{\substack{1 \leq i,j \leq 4 \\ i<j}} a_{ij}$. Clearly, $\gcd(\omega, m) = 1$. It can be verified that $\omega^{-1} = 499$. So, R's secret decryption key is

$$\left(\begin{bmatrix} 8 & 2 & 64 \\ & 128 & 32 \\ & & 16 \end{bmatrix}, 582, 499 \right).$$

R next computes B' to be

$$\begin{bmatrix} 56 & 14 & 448 \\ & 314 & 224 \\ & & 112 \end{bmatrix}.$$

Note that except for b_{23}' the b_{ij}''s tend to reveal the structure of A. So, R adds random multiples of m to each b_{ij}' to get

$$B = \begin{bmatrix} 638 & 1760 & 448 \\ & 1478 & 806 \\ & & 1276 \end{bmatrix}.$$

The above-diagonal elements of B form the public encryption key of R.

Encryption :

We assume the network convention explained in section 2.4. Let the information to be transmitted by, say, S to R be 101. Step E1 of the modified encryption procedure gives $f(1) = 1$, $f(2) = 1$, $f(3) = -1$, $f(4) = 1$. Step E2 yields

$$C = 638 - 1760 + 448 - 1478 + 806 - 1276 = -2622.$$

S transmits C to R.

Decryption :

R applies step D1 to get

$$d = \text{rem}(C\omega^{-1}, m) = \text{rem}(-2622 * 499, 582)$$
$$= \text{rem}(-1308378, 582)$$
$$= -42.$$

(We note that the test ' $|d| > \sum_{\substack{1 \leq i,j \leq 4 \\ i<j}} a_{ij}$? ' fails).

At step D2, R executes algorithm P to get

$$(A_P, A_N) = (\{64, 32, 8\}, \{128, 16, 2\}).$$

Execution of algorithm F at step D3 leads to $f(1) = 1$, $f(2) = 1$, $f(3) = -1$, $f(4) = 1$. At the end of step D4, R recovers $M = 1101$. He now discards $m(1) = 1$ to obtain 101 - the information S wanted to send him.

Example 2 :

We assume the same encryption and decryption keys as in the previous example. However, the information which S wishes to convey to R is now 011. Application of the encryption procedure produces $f(1) = 1$, $f(2) = -1$, $f(3) = 1$, $f(4) = 1$. Encryption gives

$$C = -638 + 1760 + 448 - 1478 - 806 + 1276 = 562.$$

During decryption R computes

$$d = \text{rem}(C\omega^{-1}, m) = \text{rem}(562 * 499, 582)$$
$$= \text{rem}(280438, 582)$$
$$= 496.$$

Now, ' $|d| > \sum_{\substack{1 \leq i,j \leq 4 \\ i<j}} a_{ij}$? ' is true because 496 > 250 and since $d > 0$ we replace d by $d - m = 496 - 582 = -86$.

Algorithm P yields $(A_P, A_N) = (\{64, 16, 2\}, \{128, 32, 8\})$ and algorithm F produces $f(1) = 1$, $f(2) = -1$, $f(3) = 1$, $f(4) = 1$. At the end of step D4, R obtains $M = 1011$. He discards $m(1) = 1$ and gets 011 - which is what S wanted to send him.

3. Further Comments On This System

3.1. A Variation

All along we have worked under the restriction that the above-diagonal elements of the secret matrix A be nonnegative, distinct and positive powers of 2. Thus, the elements of A can be permuted into a superincreasing sequence *. It turns out that lemma 1 and lemma 2 as well as the decryption procedure outlined above are valid if the above-diagonal elements of A satisfy the following conditions :

(i) they are distinct
(ii) they are nonnegative
(iii) they can be permuted into a superincreasing sequence.

Therefore, it is possible to build a public-key cryptosystem based on the matrix cover problem with these relaxed conditions. However, for purposes of exposition we shall confine ourselves to the 'powers of 2' version. This has another advantage, when the powers are consecutive, as we shall see in section 3.4.

3.2. Cryptocomplexity

The brute-force approach to breaking this system would involve trying out each of the possible partitions of the above-diagonal elements of the publicly-known matrix B of user R until a partition (B_P^* , B_N^*) is found such that the ciphertext $C = \sum_{b_{ij} \varepsilon B_P^*} b_{ij} - \sum_{b_{ij} \varepsilon B_N^*} b_{ij}$. (The cryptanalyst knows C).
The number of above-diagonal elements is $n' = (n^2 - n) / 2$. So, the total number of partitions of these n' elements is $n'C_0 + n'C_1 + \ldots + n'C_{n'} = 2^{n'} = 2^{(n^2-n)/2}$. However, even for $n = 15$, cryptanalysis becomes prohibitively expensive. Assuming that a partition is generated and tested every microsecond, we find that cryptanalysis would, in the worst case, take about $3 * 10^{18}$ years !

However, it is inadvisable to choose such a small value for n because a cryptanalyst could break the system by successively generating each of the 2^{15} binary messages of length n, enciphering it and comparing the resulting ciphertext with the ciphertext he has intercepted until a match is found. To render such an attack computationally infeasible, n should be about 100 or more. (Note that for $n = 100$ or more, the brute-force approach is virtually ruled out because it involves the generation and testing of 2^{4950} or more partitions).

Since the matrix cover decision problem is NP-complete, it is unlikely that an efficient (deterministic polynomial-time) algorithm will be found for it. The corresponding problem of determining the function f given that it exists, is at least as hard as the decision problem. However, it is not known whether the cryptographic problem is as hard as the hardest problem above. This is because an instance of the cryptographic problem, involving matrix B, is created from the matrix A which has certain special properties.

---

* A sequence $d_1, d_2, \ldots, d_n$ of positive integers is said to be superincreasing if $d_i > \sum_{j=1}^{i-1} d_j$, $2 \le i \le n$.

Unfortunately, complexity theory has not progressed to the point where formal proofs of secureness of public-key cryptosystems are forthcoming. Until such time we shall have to base our assessment of such systems on the failure of concerted attacks to break them.

3.3. Choice Of The a_{ij}'s

As mentioned above, in a practical system $n > 100$. We assume that $n = 100$ for all users in the system. Further, we could require that all users choose the above-diagonal elements of their secret matrix A from the set $\{1, 2, 2^2, \ldots, 2^{((100^2-100)/2 - 1)}\} = \{1, 2, 2^2, \ldots, 2^{4949}\}$. The sequence $1, 2, 2^2, \ldots, 2^{4949}$ is a minimum superincreasing sequence.* Thus, the 'powers of 2' version could prove to be advantageous because the individual elements are not too large compared to corresponding elements in some arbitrary superincreasing sequence. In terms of storage space needed to hold the private key of each user this could be useful.

Thus, the values of the elements appearing above the diagonal in A are known. A is secret in the sense that the *positions* in which individual elements occur are known only to the owner of A. The restriction of the above-diagonal elements of A to the set $\{1, 2, 2^2, \ldots, 2^{4949}\}$ does not affect, in any practical manner, the number of users who can be accommodated in this public-key system. The number of users is equal to the number of different secret matrices A, i.e., 4950 ! - an enormous number.

3.4. Storage Requirements

We consider storage requirements for a user's secret key as well as for his publicly known encryption key. A calculation of the amount of storage space required for each user's secret key (consisting of the above-diagonal elements of A and the parameters m and ω^{-1}) is given below for the case $n = 100$.

Each $a_{ij} = 2^k$ requires $(k + 1)$ bits, $0 \leq k \leq 4949$.

Together, the a_{ij}'s require $\sum_{k=0}^{4949} (k + 1) = \sum_{j=1}^{4950} j$ bits, where $j = k + 1$.

$$= (4950 * 4951) / 2 \sim 12 * 10^6 \text{ bits}$$
$$= 1.5 * 10^6 \text{ bytes,}$$

assuming 8 bits to a byte.

Let m be chosen from the range $[2^{5000}, 2^{5001}]$, (m should be chosen from a large set so that it is not found by direct search). Now, ω (and hence ω^{-1}) is less than m. So, m and ω^{-1} each require at most 5002 bits, i.e., 630 bytes. Thus, each user's secret key requires about $1.5 * 10^6 + 630 + 630$ bytes $\sim 1.5 * 10^6$ bytes.

The space needed for the a_{ij}'s could be cut down by maintaining a table such that for each element $a_{ij} = 2^k$, the subscript pair (i, j) is stored as the $(k + 1)^{\text{th}}$ entry in the table. Here $1 \leq i, j \leq 100$, $i < j$ and $0 \leq k \leq 4949$. Seven bits suffice for each subscript. The value of k itself

---

* $1, 2, 2^2, \ldots, 2^{4949}$ is a minimum superincreasing sequence in that for any other superincreasing sequence $x_0, x_1, \ldots, x_{4949}$, $2^i \leq x_i$, $i = 0, 1, 2, \ldots, 4949$. The x_i's are all positive integers.

need not be stored. Thus, each table entry requires at most 14 bits. The entire table can be accommodated using 4950 * 14 bits = 69300 bits or about 8700 bytes. An advantage of this method is that it lends itself well to the decryption operation. During decryption one requires to know the subscript pair corresponding to the largest exponent currently under consideration (algorithm P). This can be obtained easily by doing a bottom-up scan of the table. Thus, decryption is speeded up.

We were able to effect this saving in storage because all the above-diagonal elements of A were consecutive powers of 2. This could not have been achieved for any arbitrary superincreasing sequence. This is the advantage of the '(consecutive) powers of 2' version alluded to in section 3.1. Further, as already mentioned, the sequence $1, 2, 2^2, \ldots, 2^{4949}$ being a minimum superincreasing sequence there is a further saving in storage.

The storage considerations for the public-key of each user (as well as the data expansion calculation) will be simplified by assuming that no random multiples of m are added to the b_{ij}"s. Thus, $b_{ij}' = b_{ij}$, $1 \leq i,j \leq n$, $i < j$. Each b_{ij} will be an integer between 1 and $m - 1$. In the worst case each of the 4950 b_{ij}'s would require approximately 5000 bits necessitating $5000 * 4950 \sim 25 * 10^6$ bits $\sim 3 * 10^6$ bytes. Note that no saving in storage can be realized as before because the b_{ij}'s are arbitrary integers.

3.5. Data Expansion

The largest value a ciphertext C could assume is approximately $2^{5000} * 4950$. (This is a very rough estimate). This is about 2^{5013} and would need at most 5014 bits. Thus, 99 bits of information (assuming the network convention) expand into 5014 bits of ciphertext. The data expansion is thus $5014 / 99 \sim 50$.

Note that these are very conservative estimates. The presence of random multiples of m will increase the storage requirements and data expansion value drastically.

3.6. Signature Capability

Every ciphertext is an integer in the closed interval $[l, h]$. Here l and h are the smallest and largest values respectively, of $\sum_{\substack{1 \leq i,j \leq n \\ i < j}} b_{ij} f(i) f(j)$, where $B = (b_{ij})$ is the publicly known matrix under consideration. The ciphertext assumes its largest value when all the $f(i)$'s equal 1, i.e., when every bit in the message is equal to 1 (assume that the network convention holds). It is easy to show that the smallest value occurs when exactly $\lfloor n/2 \rfloor$ of the $f(i)$'s are of the same sign (for odd n this also implies that the smallest value occurs when exactly $\lfloor n/2 \rfloor$ of the $f(i)$'s are of the same sign). In all, there there 2^{n-1} distinct messages of length n (because the first bit is always 1). Thus, the solution density is equal to $2^{n-1} / (h + |l| + 1) < 2^{n-1} / h$, $(l < 0)$. For $n = 100$ and $m = 2^{5000}$, the solution density is less than 2^{-4900} - which is much, much worse than the solution density $(< 2^{-100})$ for the recommended values in the Merkle-Hellman knapsack scheme.

4. Conclusion

The system described here is remarkably similar to the Merkle-Hellman knapsack system [3] and was, in fact, motivated by that scheme. Many ideas, especially the incorporation of the trapdoor information using the parameters m and ω^{-1}, have been freely borrowed from there. In terms of signatures, data expansion and storage requirements the proposed system comes out a poor second to the Merkle-Hellman scheme.

The system proposed here is only of theoretical interest because it is also based on an NP-complete problem.

Acknowledgement

We wish to thank Dr. Dorothy Denning, Computer Sciences Department, Purdue University for several useful discussions. Part of the work reported here was supported under NSF grant MCS 80-15484.

References

1. M.R. Garey and D.S. Johnson, *Computers and Intractability : A Guide to the Theory of NP-completeness*, W.H. Freeman and Company, San Francisco, 1979.

2. D.E. Knuth, *The Art of Computer Programming : Seminumerical Algorithms*, Addison-Wesley Publishing Company, Reading, Massachusetts, 1969.

3. R.C. Merkle and M.E. Hellman, "Hiding Information and Signatures in Trapdoor Knapsacks", *IEEE Transactions on Information Theory*, Vol. IT-24, No. 5, Sept. 1978, pp. 525 - 530.

Appendix A

The proofs for lemma 1 and lemma 2 appear here.

LEMMA 1: Let A be a symmetric matrix in which all elements are nonnegative and are positive powers of 2. Also, let the above-diagonal elements be distinct. Then the Characteristic Difference d is always nonzero.

Proof: The proof proceeds by contradiction. Assume that there exists a partition (A_P , A_N) for which the Characteristic Difference is zero. Thus,

$$\sum_{a_{ij} \varepsilon A_P} a_{ij} - \sum_{a_{ij} \varepsilon A_N} a_{ij} = d = 0 .$$

Hence,

$$\sum_{a_{ij} \varepsilon A_P} a_{ij} = \sum_{a_{ij} \varepsilon A_N} a_{ij} = Y , \text{ say} .$$

Now, Y is a positive integer. All the a_{ij}'s are powers of 2. Therefore, the a_{ij}'s constituting A_P and those constituting A_N each represent Y in binary form. For instance, if A_P contains $a_{ij} = 2^k$ then in the binary representation of Y the ($k + 1$)th bit, counting right to left is 1. Otherwise it is zero. Similarly for A_N. Further, the above-diagonal elements of A, i.e., the a_{ij}'s in

$A_P \cup A_N$ are all distinct. So, $A_P \neq A_N$. This implies that Y has two distinct binary representations - an impossibility. So, our assumption that the Characteristic Difference is zero was incorrect. Hence the lemma.

It is straightforward to show that this lemma holds even if the above-diagonal elements of A are not powers of 2 but can be permuted into a superincreasing sequence. However, A should be a symmetric matrix, its elements nonnegative and the above-diagonal elements distinct.

LEMMA 2: If A is chosen as in lemma 1 then for a given Characteristic Difference d, the partition (A_P, A_N) is unique.

Proof: Again, the proof is by contradiction. Assume that there exist two distinct partitions (A_P', A_N') and (A_P'', A_N'') corresponding to a given Characteristic Difference d. Then,

$$d = \sum_{a_{ij} \varepsilon A_P'} a_{ij} - \sum_{a_{ij} \varepsilon A_N'} a_{ij}$$

$$= (2^{i_1} + 2^{i_2} + \ldots, + 2^{i_k}) - (2^{j_1} + 2^{j_2} + \ldots, + 2^{j_l}) .$$

(A.1)

Also,

$$d = \sum_{a_{ij} \varepsilon A_P''} a_{ij} - \sum_{a_{ij} \varepsilon A_N''} a_{ij}$$

$$= (2^{p_1} + 2^{p_2} + \ldots, + 2^{p_s}) - (2^{r_1} + 2^{r_2} + \ldots, + 2^{r_t})$$

(A.2)

Here the i's, j's, p's and r's are all positive integers and

$$\{i_1, i_2, \ldots, i_k, j_1, j_2, \ldots, j_l\} = \{p_1, p_2, \ldots, p_s, r_1, r_2, \ldots, r_t\} .$$

We observe that the right hand side of (A.1) can be obtained from the right hand side of (A.2) by removing certain elements from A_P'' and placing them in A_N'' and simultaneously removing certain other elements from A_N'' and placing them in A_P''. Let X_1 be the sum of the elements transferred from A_N'' to A_P'' and X_2 be the sum of the elements moved from A_P'' to A_N''. Thus, we have

$$d + 2(X_1 - X_2) = d .$$

(A.3)

So, $X_1 = X_2 = X$, say. X is a positive integer. Now, X_1 and X_2 are actually, each, the sum of some powers of 2 (by virtue of our choice of A). These powers of 2 represent X_1 and X_2 in binary form.

Since we have assumed distinct partitions (A_P', A_N') and (A_P'', A_N''), the powers of 2 constituting X_1 must differ from the powers of 2 constituting X_2 (remember that the above-diagonal elements of A are all distinct). This implies that $X = X_1 = X_2$ has two distinct binary representations - which is impossible. Thus, our original assumption regarding the existence of distinct partitions was incorrect and the assertion in the lemma follows.

The lemma holds even if the above-diagonal elements of A are not powers of 2 but can be permuted into a superincreasing sequence. Of course, A should satisfy the other requirements.

An example is constructed below, where for a given CD there are two partitions. Here the elements of the matrix are not all powers of 2 and cannot be permuted into a superincreasing sequence either. The matrix is

$$Z = \begin{bmatrix} 6 & 3 & 5 \\ & 8 & 1 \\ & & 9 \end{bmatrix}$$

It can be seen that for a Characteristic Difference $d = 2$ there are two partitions (Z_P', Z_N') and (Z_P'', Z_N'') where,

$$(Z_P', Z_N') = (\{3,5,9\}, \{1,6,8\})$$

and

$$(Z_P'', Z_N'') = (\{3,6,8\}, \{1,5,9\}).$$

Thus,

$$\sum_{z_{ij} \varepsilon Z_P'} z_{ij} - \sum_{z_{ij} \varepsilon Z_N'} z_{ij} = \sum_{z_{ij} \varepsilon Z_P''} z_{ij} - \sum_{z_{ij} \varepsilon Z_N''} z_{ij} = 2.$$

Appendix B

Our aim here is to show that the decryption procedure outlined earlier does work, i.e., it gives back the message M

It is obvious that once the partition (B_P, B_N) is found, the function f and the message M can be recovered unambiguously (assuming that the network convention holds). It is important to realize that f and hence M do not depend on the numerical values of the elements in the partition but only on their 'positional values', i.e., their subscripts.

The basic idea is to convert the 'hard' problem of finding a partition (B_P, B_N) to the 'easy' problem of finding a partition (A_P, A_N). In either case the message recovered will be the same if

$$A_P = \{ a_{ij} \mid b_{ij} \varepsilon B_P \} \text{ and } A_N = \{ a_{ij} \mid b_{ij} \varepsilon B_N \}.$$

The receiver R has at his disposal only the ciphertext C and the trapdoor parameters ω^{-1} and m to help him find (A_P, A_N).

The crux of the proof is to show that using the decryption procedure D, R finds exactly that Characteristic Difference d such that $\sum_{a_{ij} \varepsilon A_P} a_{ij} - \sum_{a_{ij} \varepsilon A_N} a_{ij} = d$, where A_P and A_N are as defined above. Lemma 2 guarantees that there is only one such partition (A_P, A_N).

We have,

$$C\omega^{-1} = (\sum_{B_P} b_{ij} - \sum_{B_N} b_{ij}) \omega^{-1}$$

$$= (\sum_{A_P} [a_{ij} \omega \bmod m] - \sum_{A_N} [a_{ij} \omega \bmod m]) \omega^{-1}$$

$$= (\{\sum_{A_P} [a_{ij} \omega \bmod m]\} \omega^{-1} - \{\sum_{A_N} [a_{ij} \omega \bmod m]\} \omega^{-1})$$

$$= (\{\sum_{A_P}[\,a_{ij}\omega - c_{ij}m\,]\,\}\omega^{-1} - \{\sum_{A_N}[\,a_{ij}\omega - c_{ij}m\,]\,\}\omega^{-1})$$

$$= (\sum_{A_P}[\,a_{ij}\omega\omega^{-1} - c_{ij}m\omega^{-1}\,] - \sum_{A_N}[\,a_{ij}\omega\omega^{-1} - c_{ij}m\omega^{-1}\,])$$

$$= (\sum_{A_P}[\,a_{ij} + a_{ij}\alpha m - c_{ij}m\omega^{-1}\,] - \sum_{A_N}[\,a_{ij} + a_{ij}\alpha m - c_{ij}m\omega^{-1}\,])$$

$$= ([\sum_{A_P} a_{ij} - \sum_{A_N} a_{ij}\,] + m[\,\{\sum_{A_P}[\,a_{ij}\alpha - c_{ij}\omega^{-1}\,]\,\} - \{\sum_{A_N}[\,a_{ij}\alpha - c_{ij}\omega^{-1}\,]\,\}\,])$$

$$= ([\sum_{A_P} a_{ij} - \sum_{A_N} a_{ij}\,] + km),$$

where $k = \sum_{A_P}[\,a_{ij}\alpha - c_{ij}\omega^{-1}\,] - \sum_{A_N}[\,a_{ij}\alpha - c_{ij}\omega^{-1}\,]$

So,
$$\text{rem}(C\omega^{-1}, m) = \text{rem}([\,\{\sum_{A_P} a_{ij} - \sum_{A_N} a_{ij}\} + km\,], m) \qquad (B.1)$$

At this point two cases must be considered:

Case (i): $\text{rem}(C\omega^{-1}, m) > 0$ and **Case (ii):** $\text{rem}(C\omega^{-1}, m) < 0$.

(Note that $\text{rem}(C\omega^{-1}, m) = 0$ is not possible for, from (B.1) we see that this would imply that $(\sum_{A_P} a_{ij} - \sum_{A_N} a_{ij})$ be an integral multiple of m or be equal to zero. But $m > \sum_{A_P \cup A_N} a_{ij}$ and $(\sum_{A_P} a_{ij} - \sum_{A_N} a_{ij}) \leq \sum_{A_P \cup A_N} a_{ij}$. Also, $(\sum_{A_P} a_{ij} - \sum_{A_N} a_{ij}) \neq 0$ follows from lemma 1.)

Case (i): $\text{rem}(C\omega^{-1}, m) > 0$.

The right hand side of (B.1) must also be > 0. Otherwise a mathematical contradiction arises. The sign of the right hand side depends on the sign of $(\sum_{A_P} a_{ij} - \sum_{A_N} a_{ij})$. If $(\sum_{A_P} a_{ij} - \sum_{A_N} a_{ij}) > 0$ then from (B.1) we have,

$$\text{rem}(C\omega^{-1}, m) = \text{rem}([\sum_{A_P} a_{ij} - \sum_{A_N} a_{ij}\,], m) = (\sum_{A_P} a_{ij} - \sum_{A_N} a_{ij}),$$

because $m > (\sum_{A_P} a_{ij} - \sum_{A_N} a_{ij})$.

$$\leq \sum_{A_P \cup A_N} a_{ij}.$$

If $(\sum_{A_P} a_{ij} - \sum_{A_N} a_{ij}) < 0$ then from (B.1) we have,

$$\text{rem}(C\omega^{-1}, m) = \text{rem}([\{\sum_{A_P} a_{ij} - \sum_{A_N} a_{ij} + m\} + (k-1)m], m)$$

$$= \text{rem}([\sum_{A_P} a_{ij} - \sum_{A_N} a_{ij} + m], m)$$

$$= (\sum_{A_P} a_{ij} - \sum_{A_N} a_{ij}) + m \quad \text{because } m > \sum_{A_P} a_{ij} - \sum_{A_N} a_{ij} + m.$$

$$> \sum_{A_P \cup A_N} a_{ij} \quad \text{since } m > 2 \sum_{A_P \cup A_N} a_{ij}.$$

Thus, when $\text{rem}(C\omega^{-1}, m) > 0$ we have,

$$\text{rem}(C\omega^{-1}, m) = \sum_{A_P} a_{ij} - \sum_{A_N} a_{ij} \leq \sum_{A_P \cup A_N} a_{ij}$$

or

$$\text{rem}(C\omega^{-1}, m) = \sum_{A_P} a_{ij} - \sum_{A_N} a_{ij} + m > \sum_{A_P \cup A_N} a_{ij}.$$

Which of the two actually holds is determined in step D1 of the decryption procedure by first setting d equal to $\text{rem}(C\omega^{-1}, m)$ and then comparing $|d| = d$ with $\sum_{A_P \cup A_N} a_{ij}$. If $d > \sum_{A_P \cup A_N} a_{ij}$ then it is replaced by $d - m = (\sum_{A_P} a_{ij} - \sum_{A_N} a_{ij})$. Note that the choice of m (m *must* be greater than $2 \sum_{A_P \cup A_N} a_{ij}$) is crucial because it helps to 'show up' the right value of d.

Case (ii) : $\text{rem}(C\omega^{-1}, m) < 0$.

The right hand side of (B.1) must also be < 0 in order to avoid a mathematical absurdity. The sign of the right hand side is determined by the sign of $(\sum_{A_P} a_{ij} - \sum_{A_N} a_{ij})$. If $(\sum_{A_P} a_{ij} - \sum_{A_N} a_{ij}) < 0$ then from (B.1) we have,

$$\text{rem}(C\omega^{-1}, m) = \text{rem}([\sum_{A_P} a_{ij} - \sum_{A_N} a_{ij}], m) = (\sum_{A_P} a_{ij} - \sum_{A_N} a_{ij}),$$

because $m > (\sum_{A_P} a_{ij} - \sum_{A_N} a_{ij})$.

So, $|\text{rem}(C\omega^{-1}, m)| \leq \sum_{A_P \cup A_N} a_{ij}$.

If $(\sum_{A_P} a_{ij} - \sum_{A_N} a_{ij}) > 0$ then from (B.1) we have,

$$\text{rem}(C\omega^{-1}, m) = \text{rem}([\{\sum_{A_P} a_{ij} - \sum_{A_N} a_{ij} - m\} + (k+1)m], m)$$

$$= \text{rem}([\sum_{A_P} a_{ij} - \sum_{A_N} a_{ij} - m], m)$$

$$= (\sum_{A_P} a_{ij} - \sum_{A_N} a_{ij}) - m \quad \text{because } m > \sum_{A_P} a_{ij} - \sum_{A_N} a_{ij} - m.$$

So,
$$|\operatorname{rem}(C\omega^{-1}, m)| > \sum_{A_P \cup A_N} a_{ij} \quad \text{because } m > 2 \sum_{A_P \cup A_N} a_{ij}.$$

Thus, when $\operatorname{rem}(C\omega^{-1}, m) < 0$ we have,
$$|\operatorname{rem}(C\omega^{-1}, m)| = |\sum_{A_P} a_{ij} - \sum_{A_N} a_{ij}| \le \sum_{A_P \cup A_N} a_{ij}$$

or

$$|\operatorname{rem}(C\omega^{-1}, m)| = |\sum_{A_P} a_{ij} - \sum_{A_N} a_{ij} - m| > \sum_{A_P \cup A_N} a_{ij}.$$

Again, to determine which of the two above applies, step D1 in the decryption procedure sets d equal to $\operatorname{rem}(C\omega^{-1}, m)$ and compares $|d|$ with $\sum_{A_P \cup A_N} a_{ij}$. If $|d| > \sum_{A_P \cup A_N} a_{ij}$ then d is replaced by $d + m = (\sum_{A_P} a_{ij} - \sum_{A_N} a_{ij})$. Note again the reason for our choice of m as $m > 2 \sum_{A_P \cup A_N} a_{ij}$. In general m could be $\beta \sum_{A_P \cup A_N} a_{ij}$, where $\beta \ge 2$.

The above proof did not consider the addition of random multiples of m to the above-diagonal elements of the matrix B' to obtain B. This can easily be incorporated. In any case these random multiples would get 'swallowed up' in the term 'km' in (B.1) above. Thereafter the proof would proceed as before.

INFINITE STRUCTURES IN INFORMATION THEORY

G. R. Blakley

Department of Mathematics
Texas A&M University
College Station, TX 77843

Laif Swanson

Communications Systems
Research Section
Jet Propulsion Laboratory of
California Institute of
Technology
Pasadena, CA 91103

The idea of infinite structures, even of continua, in information theory is not a new one [KO56]. This note is devoted to infinite one-time pads [BL81] and an infinite projective geometric [BL79] threshold scheme. Perhaps an infinite structure can be better understood than its finite analog if it is amenable to investigation by methods from calculus or harmonic analysis. It is conceivable that an existing error control code, pool/split/restitute process [AS82], or cryptosystem can be better understood by examining an infinite version.

1. INFINITE ONE-TIME PADS

The classical one-time pad works as follows. An n-bit word p (i.e. a member of $GF(2^n)$) is chosen at random and each of two people takes a copy of p. When one of the people has an n-bit message s to send to the other, he forms the bitwise exclusive or, $p + s$, of the pad and the message and transmits $p + s$ to the other party. The receiver merely forms $p + p + s = s$ and, thus, has recovered the message. The important feature of the pad is its Shannon perfect security. [SH49, pp. 679-683] An opponent of the sender and receiver cannot modify a, perhaps shrewd, initial guess concerning s if he intercepts $p + s$. In other words, for every t belonging to $GF(2^n)$, the equality

Probability (the message is t) =

Probability (the message is t | given that p + s is intercepted).

It is also true that

Probability (the message is t) =

Probability (the message is t | given that p is somehow known)

A priori probabilities are the same as a posteriori probabilities, in other words. You need both p and p + s to find s. Knowing just one of them is knowing nothing about s you didn't already know just before the sender composed the message.

Is it possible to retain Shannon perfect security when some messages s are infinitely more probable than others? Let us make this explicit by an example. A wind sock may hang limp if the wind is less than 1 knot. For larger wind speeds it points some direction. It is quite plausible, then, that the message

No measurable wind

has probability 30 %, since the windsock is limp a fair amount of the time. No single wind direction has positive probability, but for each interval there is positive probability that the wind direction lies in that interval. It may be that the probability that the wind lies between N and W is 45% whereas the probability that it lies between S and SE is only 4%. At any rate there is some density function on the interval [0°, 360°) whose integral is 0.7. The corresponding probability measure has a lump (i.e. an atom) with measure 0.3 and a smooth stretch (an interval on which it is absolutely continuous [BA72, p. 84] with respect to ordinary Lebesgue measure) with measure 0.7. In this sense the message

No measurable wind

is infinitely more probable (since its probability is 0.3) than the message

Wind NNE

which has probability 0.

Is it possible to build a perfectly secure one-time pad when the a priori probabilities are this divergent? Yes. We will call such pads raisin pudding pads because their a priori probability measures have some lumps and some smooth parts.

A raisin pudding pad for the wind example is as follows. The collection of messages will be

$$M = \{a\} \cup [0,1)$$

and the probability measure μ will be induced by the requirements

$$\mu(a) > 0$$

$$\mu(\{a\} \cup [0,1)) = 1$$

μ is absolutely continuous with respect to ordinary Lebesgue measure on $[0,1)$.

For the pad p choose any (infinite) sequence

$$p = \{p(1), p(2), p(3),...\}$$

of bits (zeros and ones) at random. In other words let p be a uniformly distributed random variable in $[0,1]$. Now choose any message m belonging to M. We define the pad encoding of m to be the infinite sequence

$$e = \{e(1), e(2), e(3),...\}$$

of bits formed as follows. If $m = a$ then

$$e(1) = p(1)$$

$$e(k) = 1 - p(k), \qquad k \geq 2$$

If $m = \{m(1), m(2), m(3),...\}$ belongs to $[0,1)$ then

$$e(t) = 1 - p(t), \qquad t \text{ odd}$$

$$e(2k) = p(2k) + m(k) \qquad (\text{mod } 2).$$

Clearly this is Shannon perfectly secure. Decoding is easy. The message m is a if

$$e(1) + p(1) = 0 \qquad (\text{mod } 2).$$

The message m belongs to $[0,1)$ if

$$e(1) + p(1) = 1 \qquad (\text{mod } 2)$$

in which case

$$m(i) = p(2i) + e(2i) \qquad (\mod 2)$$

for every positive integer i.

It is evident, now, how to construct a raisin-pudding one-time pad for a message space of the form $M = A \cup [0,1]$, where A is finite or countably infinite, and:

$\mu(a) > 0$ for each a belonging to A;

$\Sigma\mu(a) + \mu([0,1)) = 1$, where the sum is over a belonging to A;

μ is absolutely continuous with respect to ordinary Lebesgue measure on $[0,1)$.

Enumerate A in some fashion

$$A = \{a(1), a(2), a(3), \ldots\}$$

Choose the bit sequence $p = \{p(1), p(2), p(3), \ldots\}$ at random from $[0,1]$. Choose m belonging to M. If m belongs to A then $m = a(j)$ for some j, so let

$e(i) = p(i)$, $\qquad 1 \leq i \leq j$

$e(k) = 1 - p(i)$, $\qquad j + 1 \leq k$

If $m = \{m(1), m(2), m(3), \ldots\}$ belongs to $[0,1)$ let

$e(t) = 1 - p(t)$, $\qquad$ t odd

$e(2s) = m(s) + p(2s)$ $\qquad (\mod 2)$

Evidently, for every t which belongs to M

Probability (message m equals t)

= Probability (message m equals t | given that p is known)

= Probability (message m equals t | given that e is known),

because of the fact that p was chosen at random, and of the way e was constructed from m and p. The decoding process is trivial. The receiver adds p to e modulo 2 in each entry to get

$$d = \{d(1), d(2), d(3), \ldots\}.$$

If $d(1) = 0$ the message m belongs to A, and in fact is $a(j)$, given that

$$d(1) = d(2) = \ldots = d(j) = 0$$

and

$$1 = d(j+1) = d(j+2) = \ldots .$$

If $d(1) = 1$ the message belongs to $[0,1)$, and in fact is

$$m = \{m(1), m(2), m(3), \ldots\} = \{e(2)+p(2), e(4)+p(4), e(6)+p(6), \ldots\}$$

gotten by adding (modulo 2) the even entries of the e and p sequences.

The raisin pudding one time pad has a feature in common with the classical one-time pad. The set of possible pads p is in one to one correspondence with the set of possible messages m. But our next example is of a one-time pad with uncountably many possible pads p but only countably many messages m.

For this example let $A = \{a(1), a(2), a(3), \ldots\}$ be a countably infinite collection of messages. And let μ be a probability measure on A which is strictly positive at each point of A. Let

$$w(i) = \sum_{j=1}^{i} \mu(a(j))$$

for every positive integer i. By the usual convention $w(0) = 0$. Let

$$L(i) = [w(i-1), w(i))$$

for every positive integer i. The half open intervals $L(i)$ are pairwise disjoint and their union is the half open unit interval $[0,1)$. The pad $p = \{p(1), p(2), p(3), \ldots\}$ is a random sequence of bits (i.e. a random variable uniformly distributed in $[0,1)$). A message m is a point $a(t)$ of A. To encode m using p choose a bit sequence

$$y = \{y(0), y(1), y(2), \ldots\}$$

at random from $L(t)$ and form the bit sequence

$$e = \{e(1), e(2), e(3), \ldots\}$$

by setting

$$e(j) = p(j) + y(j) \qquad (\text{mod } 2)$$

for every positive integer j. It is clearly Shannon perfectly secure. It is easy to decode by finding

$$y(j) = e(j) + p(j) \qquad (\text{mod } 2)$$

This tells the value of y, which can be found to lie in a unique $L(t)$. So the message m is equal to $a(t)$.

2. THRESHOLD SCHEMES

Threshold schemes are now widely publicized in standard texts [DE82]. But, because they are still relatively new we repeat the basic properties of a k out of n threshold scheme. A piece of information (i.e. a number) called a substance is to be safeguarded by using a lot of garbage (other random numbers), together with the substance, to produce n shadows (mathematical objects, usually other numbers in the same size range as the substance) in such a way that it is computationally trivially to reconstruct the substance from any k shadows, but utterly impossible to get an idea of the substance from a knowledge of only $k - 1$ shadows. It is this last statement which must be made precise.

In a Bloom [BL81b] threshold scheme or a Shamir [SH79] scheme it amounts to Shannon perfect security [BL80]. To be specific a set S is chosen in advance (usually a set of b-bit words, i.e. nonnegative integers smaller than 2^b). Substance and shadows are all chosen from S. In a Bloom or a Shamir k out of n scheme one needs $k - 1$ pieces of garbage ($k - 1$ b-bit random words) as well as a substance to produce the needed n shadows. Shannon perfect security is the statement that, for every x belonging to S,

A priori probability (substance = x) =

Conditional probability (substance = x | given knowledge of any $k - 1$ shadows).

Evidently the one-time pad is the 2 out of 2 case of some threshold scheme. It happens that it is a case of the Bloom fast threshold scheme.

The Bloom scheme and the Shamir scheme are <u>rigid</u> in the sense that once you have k shadows in a k out of n scheme you know all the other shadows, as well as the substance. For this reason either of them is easy to turn them into a pool/split/restitute process [AS82]. We will give an infinite version of the Blakley geometric threshold scheme [BL79]. It is not rigid. Knowledge of k shadows in its k out of n case yields the substance, but not the other shadows. Thus there is no obvious way to turn it into a pool/split/restitute process.

The idea behind a geometric threshold scheme is easy to visualize. Consider a 2 out of n scheme. A substance m (message, if you will) is a point from some field. Pick another number r at random to get a pair (m,r) which can be regarded as a point in the plane. A shadow is just a randomly chosen line lying on this point (and not parallel to either coordinate axis). One shadow yields no information regarding the point. But two shadows intersect at the point, and this reveals m. To get a 3 out of n scheme make the substance m the first coordinate of a point (m,r,s) in 3 dimensional space and let the shadows be randomly oriented planes on the point (m,r,s). Knowledge of only 2 planes gives no information about m. But any 3 planes intersect only at (m,r,s), and thus yield n readily.

Similarly, for k out of n schemes, the substance becomes the first coordinate of a point in a k dimensional space, and the shadows are randomly chosen hyperplanes (subspaces of codimension 1) on the point.

Whether this can be made rigorous or not depends on a few points we have glossed over. Is the space a vector space, an affine space or a projective space? Is the field from which the substance m is selected finite or infinite? If the field is finite it is possible to produce geometric schemes in any one of the three kinds of spaces. These schemes are all Shannon almost perfectly secure [SH49, pp. 679-683] in the sense that finding k - 1 shadows in a k out of n scheme does not enable an oponent to modify any a priori guess as to probabilities by a factor of 2. For every message x

[A Priori Probability (x is the substance)]/2

    < Conditional Probability (x is the substance | given knowledge of k - 1 shadows)

    < 2[A Priori Probability (x is the substance)]

in a k out of n scheme.

As our earlier examples with one time pads indicated, the way to get Shannon perfect security or Shannon almost perfect security in the k out of n case of a threshold scheme is to choose some shadows as uniformly distributed random variables on an appropriate probability space.

Vector spaces, affine spaces and projective spaces over a finite field have only finitely many hyperplanes, so a natural uniform probability measure, the counting measure, is available. But vector spaces and affine spaces over infinite fields do not readily admit of such uniform measures. Projective spaces are better behaved in this regard. They (like vector spaces and affine spaces) are homogeneous. But they, unlike the others, are also compact. And homogeneous compact structures admit of a natural uniform measure, the "volume" measure.

Real and complex projective spaces are compact [MC67, p. 141] and homogeneous [WA71, p. 128], and thus admit of a "volume" measure. In fact complex projective spaces have very well behaved Grassmanians [GR78, pp. 193-211]. The collection of all hyperplanes in projective k dimensional space is a Grassmanian. But we can get by with the observation that points are in one to one correspondence with hyperplanes by duality [YA68, pp. 255-257]. Hence the homogeneity and compactness of the manifold of points induces homogeneity and compactness on the (Grassman) manifold of hyperplanes. This enables us to put the "volume" measure on the set of all hyperplanes and to speak of choice of a hyperplane as being a uniformly distributed random variable on the set of all hyperplanes.

3. A REAL PROJECTIVE GEOMETRIC THRESHOLD SCHEME

With this background we describe the k out of n real projective geometric threshold scheme. Those who prefer complex projective spaces can replace the field R by C everywhere below. The other modifications are easy to see.

The points of real k dimensional projective space $P(k)$ are the one dimensional vector subspaces of a real $k + 1$ dimensional vector space $V(k + 1)$. More generally b dimensional projective subspaces of $P(k)$ are $b + 1$ dimensional vector subspaces of $V(k + 1)$. So hyperplanes of $P(k)$ are hyperplanes of $V(k + 1)$ which contain the origin.

We start with a set S of Lebesgue measure zero in the open unit interval $(0,1)$. S is the set of substances (messages). On S there can be any measure you choose. In other words some substances can be more likely than others, but it is easy to tell when you have a substance (highly unlikely) or just a randomly chosen number from (0,1). The substance (message) is a real number s such that 0 < s < 1. Now the projective point π corresponding to s in a k out of n threshold scheme is gotten as follows. Choose any i belonging to $\{1, 2, \ldots, k + 1\}$. Pick a point p belonging to V(k + 1) by setting $p(i) = s/\sqrt{k}$ and by choosing p(1),...,p(i-1), p(i+1),...,p(k+1) so that

$$p(1)^2 + \ldots + p(i-1)^2 + p(i+1)^2 + \ldots + p(k+1)^2 = 1,$$

in other words, by choosing a point at random from the unit sphere of V(k) (here we assume "at random" refers to a uniform probability distribution on this unit sphere).

The element π belonging to P(k) corresponding to s is defined to be the unique one dimensional vector subspace of V(k + 1) containing p. The n shadows of s are n randomly chosen hyperplanes of P(k), all of which contain π. Another way to say this is to say that a shadow is a k dimensional vector subspace of V(k + 1) which contains p. Again, the notion of "random choice" is possible because there is a uniform distribution on the points of P(k) and, hence by duality, on the hyperplanes of P(k). We choose to reject a point on the unit sphere in V(k) if it has a zero coordinate. Similarly we reject a hyperplane in P(k) if it is of the form, considered as a k dimensional vector subspace of V(k + 1),

$$\left\{ x : \sum_{i=1}^{k+1} a(i)x(i) = 0 \right\}$$

where some a(i) is zero. These are both probability zero occurrences, and so a rejection will seldom occur and will not skew the probabilities when it does occur. A further probability zero occurrence is also intolerable, the possibility that there are k of the n shadows (projective hyperplanes) whose intersection contains more than the point π. Once again, a simple test can detect this and replace one or more projective hyperplanes so the problem disappears.

Somebody who has k shadows can intersect them to find the projective point π belonging to P(k), i.e. the one dimensional vector subspace of V(k + 1) within which p lies. At this point there are only k + 1 alternatives for the value of p, since the sum of the squares of some k of its entries must be 1. So, if q is a nonzero vector belonging to π, then s must be of the form

$$k^{1/2} |q(i)| (q(1)^2 + \ldots + q(i-1)^2 + q(i+1)^2 + \ldots + q(k+1)^2)^{-1/2}$$

for some i belonging to $\{1, 2, \ldots, k+1\}$. With probability 1, one of these numbers lies in the set of possible substances and the remaining k do not. So recovery of s is complete. Note that the probability 1 occurrence that only one candidate lies in the set of substances can actually be made a certainty at the outset. This is done by simply rejecting a point on the unit sphere of $V(k)$ to represent s, if an examination of all candidates determined by π turns up another possible substance besides s among them.

The arguments which establish Shannon almost perfect security are like those in [BL79]. They rely on the uniform probability measures on the unit sphere of $V(k)$ and on the collection of all hyperplanes in $P(k)$ which contain π (This, by duality, can be identified with all points in some fixed hyperplane of $P(k)$, i.e. with $P(k-1)$, which we have seen above admits of a uniform probability measure). The probability measure on the set S of substances can, of course, be arbitrary, though S must be of Lebesgue measure zero in $(0,1)$.

4. DISCUSSION

It would be interesting to have entire Bloom threshold schemes which have the one-time pads of this paper as special cases. It would, in fact, be nice to have any example of an infinite rigid threshold scheme, so that we could construct infinite pool/split/restitute processes. Good infinite cryptosystems would have obvious applicability to secure analog voice transmission. The infinite Caesar cipher with key k defined by the function

$$c_k(z) = kz$$

from the unit circle $U = \{z: |z| = 1\}$ in the complex plane to itself, where k belongs to U, does not seem to qualify for the adjective "good". It might prove useful to examine infinite analogs of various other well-known cryptosystems.

The actual implementation of infinite structures in information theory would involve considerable practical difficulty which we have neglected here. We content ourselves by observing that continuity problems might look different in some p-adic norm, rather than the standard "distance" norm on the set R of real numbers, and that some nominally infinite structure might be implemented in truncated finite versions whose size could be varied to suit the occasion.

REFERENCES

AS82   C. A. Asmuth and G. R. Blakley, Pooling, splitting and restituting information to overcome total failure of some channels of communication, Proceedings of the 1982 Symposium on Security and Privacy, IEEE Computer Society, Los Angeles, California (1982), pp. 156-169.

BA72   H. Bauer, Probability Theory and Elements of Measure Theory, Holt, Rinehart and Winston, New York (1969).

BL79   G. R. Blakley, Safeguarding cryptographic keys, Proceedings of the National Computer Conference, 1979, American Federation of Information Processing Societies-Conference Proceedings, Vol. 48 (1979), pp. 313-317.

BL80   G. R. Blakley, One-time pads are key safeguarding schemes, not cryptosystems. Fast key safeguarding schemes (threshold schemes) exist, Proceedings of the 1980 Symposium on Security and Privacy, IEEE Computer Society, Long Beach, California (1980), pp. 108-113.

BL81   G. R. Blakley and Laif Swanson, Security proofs for information protection systems, Proceedings of the 1981 Symposium on Security and Privacy, IEEE Computer Society (1981), Los Angeles, California, pp. 75-82.

BL81b   J. Bloom, A note on superfast threshold schemes, Preprint, Texas A&M University, Department of Mathematics (1981),
and
Threshold schemes and error correcting codes, Astracts of Papers Presented to the American Mathematical Society, Vol. 2 (1981), p. 230.

DE82   D. E. R. Denning, Cryptography and Data Security, Addison Wesley, Reading, Massachusetts (1982).

GR78   P. Griffiths and J. Harris, Principles of Algebraic Geometry, Wiley-Interscience, New York (1978).

KO56   A. N. Kolmogorov, On the Shannon theory of information in the case of continuous signals, IEEE Transactions on Information Theory, Vol. IT-2 (1956), pp. 102-108,
Reprinted as pages 238-244 in
D. Slepian (editor), Key Papers in the Development of Information Theory, IEEE Press, New York (1974).

MC67   G. McCarty, Topology: An Introduction with Application to Topological Groups, McGraw-Hill, New York (1967).

SH49   C. E. Shannon, Communication theory of secrecy systems, Bell System Technical Journal, Vol. 28 (1949), pp. 656-715.
SH79   A. Shamir, How to share a secret, Communications of the ACM, vol. 22 (1979), pp. 612-613.
WA71   F. W. Warner, Foundations of Differentiable Manifolds and Lie Groups, Springer-Verlag, New York (1971).
YA68   P. B. Yale, Geometry and Symmetry, Holden-Day, San Francisco (1968).

A FAST MODULAR MULTIPLICATION ALGORITHM WITH

APPLICATION TO TWO KEY CRYPTOGRAPHY[*]

Ernest F. Brickell

Sandia National Laboratories
Albuquerque, New Mexico 87185

ABSTRACT

This paper presents an algorithm which will perform multiplication modulo C in $\lceil \log_2 C \rceil + 7$ clock pulses.

INTRODUCTION

The 1976 paper of Diffie and Hellman [1] on two key cryptology started a search for two key cryptosystems. Several that have been discovered [3,5,6] depend on multiplication or exponentiation using modular arithmetic. However, to obtain reasonable cryptosecurity, the moduli must often be hundreds of bits long. Using conventional modular multiplication routines, the two key systems are several orders of magnitude slower than one key cryptosystems offering similar cryptosecurity, such as the Data Encryption Standard, DES [7]. Therefore, it is important to find algorithms which will speed up modular multiplication.

If n is the length of the modulus, a conventional modular multiplication algorithm will multiply in $O(n^2)$ clock pulses. In 1981, Norris and Simmons [4] developed an algorithm which will multiply in 2n clock pulses. The algorithm we will present here will multiply in n+7 clock pulses.

---

[*] This work performed at Sandia National Laboratories supported by the U. S. Department of Energy under contract number DE-AC04-76DP00789.

TERMINOLOGY

If X is an n-bit register we will use x_i to represent the ith bit of the register contents, i.e., X contains the binary number $\sum_{i=0}^{n-1} x_i 2^i$.

Robertson, Rohatsche, and Wallace [8,9,10] developed the notion of a carry-save adder, which was used in the ILLIAC4. In a carry-save adder, the accumulator consists of a sum register S and a carry register C. To add a binary number B to the accumulator, one computes, in one clock pulse

$$s_i \leftarrow s_i \oplus c_i \oplus b_i$$

and

$$c_{i+1} \leftarrow s_i c_i \vee s_i b_i \vee c_i b_i$$

for all i. Since the carries of each operation are saved in the C-register, there is no carry propagation. Thus the adders are very fast.

Norris and Simmons developed a modification of this, which they called a delayed carry adder. The accumulator still consists of two registers, T and D. To add a binary number B to the accumulator, one forms, in one clock pulse,

$$s_i \leftarrow t_i \oplus d_i \oplus b_i$$
$$c_{i+1} \leftarrow t_i d_i \vee t_i b_i \vee b_i d_i$$

and then

$$t_i \leftarrow s_i \oplus c_i$$
$$d_{i+1} \leftarrow s_i c_i$$

for all i. Note that $d_{i+1} t_i = 0$ for all i, and that $d_0 = 0$.

Thus we define a delayed carry integer of length n to be an ordered pair of registers $A = (a, \alpha)$ such that a and α are n-bit registers, $a_{i-1} \alpha_i = 0$ for all i, $1 \leq i \leq n-1$, and $\alpha_0 = 0$. Also, A is assumed to represent the positive integer $\sum_{i=0}^{n-1} (a_i + \alpha_i) 2^i$. When we refer to the high order j bits of A, we mean the high order j bits of both a and α, i.e., a_i, α_i for $i \geq n-j$. Figure 1 is the diagram we use for a half adder with input bits x,y and output bits d,δ. Notice that the output bits satisfy $d\delta = 0$.

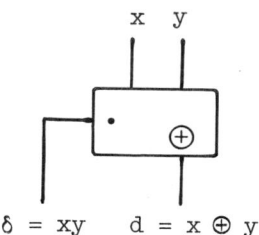

$\delta = xy \quad d = x \oplus y$

Figure 1. Half Adder

MODULAR MULTIPLICATION USING DELAYED CARRY NUMBERS

Let A and B be two delayed carry integers of length n. The product $D = A \cdot B$ can be computed by executing the following n steps.

$$D \leftarrow a_0 \cdot B + a_1 \cdot 2B$$

$$D \leftarrow D + a_1 \cdot 2B + a_2 \cdot 2^2 B$$

$$\vdots$$

$$D \leftarrow D + a_{n-2} \cdot 2^{n-2} B + a_{n-1} \cdot 2^{n-1} B$$

$$D \leftarrow D + a_{n-1} \cdot 2^{n-1} B$$

Since $a_i a_{i+1} = 0$, either $a_i \cdot 2^i B$ or $a_{i+1} \cdot 2^{i+1} B$ is 0. Each step requires only a shift of B and the addition of at most two delayed carry integers. One reason for using delayed carry rather than carry save should now be apparent. Using carry save, it is possible to have $a_i a_{i+1} = 1$. Then step i+1 would require the addition of three carry-save numbers, namely D, $2^i B$, and $2^{i+1} B$.

After executing the preceding n steps, $D = A \cdot B$ and D is a delayed carry integer of length 2n. We can find $E \equiv D \bmod C$ where $0 \leq E < C$ by the following procedure, which is essentially a division routine in which the only output is the remainder.

$$\text{If } (D \geq 2^{n-1} C) \text{ then } D \leftarrow D - 2^{n-1} C$$

$$\text{If } (D \geq 2^{n-2} C) \text{ then } D \leftarrow D - 2^{n-2} C$$

$$\vdots$$

$$\text{If } (D \geq C) \text{ then } D \leftarrow D - C$$

$$E \leftarrow D$$

There is, however, a method for combining the multiplication procedure with the division procedure. We can reverse the order of the additions in the multiplication procedure.

$$D \leftarrow a_{n-1} \cdot 2^{n-1} B$$

$$D \leftarrow D + a_{n-2} \cdot 2^{n-2} B + \alpha_{n-1} \cdot 2^{n-1} B$$

$$\cdot$$
$$\cdot$$
$$\cdot$$

$$D \leftarrow D + a_0 \cdot B + \alpha_1 \cdot 2B$$

After a few steps, the top (high-order) bits of D are not changing very often. It might be possible to determine if a given multiple of C, say $2^\ell C$, should be subtracted from D before the multiplication procedure has finished many steps. In examining this idea closely, there are two problems that arise. $D-2^\ell C$ might be negative after step j, but positive after step i, for some $i > j$. However a more serious problem is the time necessary to determine the sign of $D-2^\ell C$. When using delayed carry or carry-save arithmetic, magnitude comparison is always much slower than addition. To determine the sign of $D-2^\ell C$, we must find $E = D-2^\ell C$ in the redundant number system and then essentially convert E to binary.

Fortunately, there is a solution to these problems. At the appropriate step, allow the subtraction of either $2^\ell C$ or $2^{\ell-1} C$ from D.

To explain this cryptic solution, we must present more detail. We start subtracting multiples of C eleven steps after the multiplication procedure is started. So D, the accumulator, is a delayed carry register of length $n + 11$. Assume that $K = 2^n - C$ is stored as a binary number (i.e., K is stored in only one register). t_1 and t_2 are two bits used as control bits. t_2 and t_1 are determined in essentially, but not precisely, the following manner. t_2 is set to 1 when the top 4 bits of D are greater than the top 4 bits of $2^{11} C$. t_1 is set to 1 when $t_2 = 0$ and the top 4 bits of D are greater than the top 3 bits of $2^{10} C$.

ALGORITHM M

$D \leftarrow 0$, $t_1 \leftarrow 0$, $t_2 \leftarrow 0$
Do $j = 1$ to $n + 10$
    $B^* \leftarrow a_{n-1} B + \alpha_n 2B$
    $K^* \leftarrow t_2 2^{11} K + t_1 2^{10} K$
    $D \leftarrow 2(D + B^* + K^*)$
    $A \leftarrow 2A$
    Add the top 4 bits of D to the top 4 bits of $2^{11}K$.
    If there is an overflow of 2^{11}, then $t_2 \leftarrow 1$,
        else $t_2 \leftarrow 0$.
    Add the top 4 bits of D to the top 3 bits of $2^{10}K$.
    If there is an overflow of 2^{10} and $t^2 = 0$,
        then $t_1 \leftarrow 1$
        else $t_1 \leftarrow 0$.
End.

Delayed carry register A actually contains n bits for register a, and n+1 bits for register α so that α_n can be stored after the left shift of A ($A \leftarrow 2A$).

Figure 2 shows that the statement $D \leftarrow 2(D + B^* + K^*)$ can be computed with a cascade of 5 half adders. However for $i > n$, b_i^* and β_i^* are 0. So with a cascade of 5 half adders, we can compute the top bits of $D \leftarrow 2(D + B^* + K^*)$ with the adder shown in Figure 3 and also compute t_2 and t_1.

Since D is a delayed carry register of length $n + 11$, the bits in Figure 2 marked δ_{n+11}, d_{n+11}, and δ_{n+12} are the overflow from D. Let D^j denote the contents of register D at the end of the jth step. Then $D^j \leftarrow 2(D^{j-1} + B^* + K^*)$ can be written more precisely as

$$(2\delta_{n+12} + d_{n+11} + \delta_{n+11})2^{n+11} + D^j$$
$$= 2(D^{j-1} + t_2 2^{11}K + t_1 2^{10}K + a_{n-1}B + \alpha_n 2B) \ .$$

We have

$$(2t_2 + t_1)2^{11}K = (2t_2 + t_1)2^{n+11} - (2t_2 + t_1)2^{11}C \ .$$

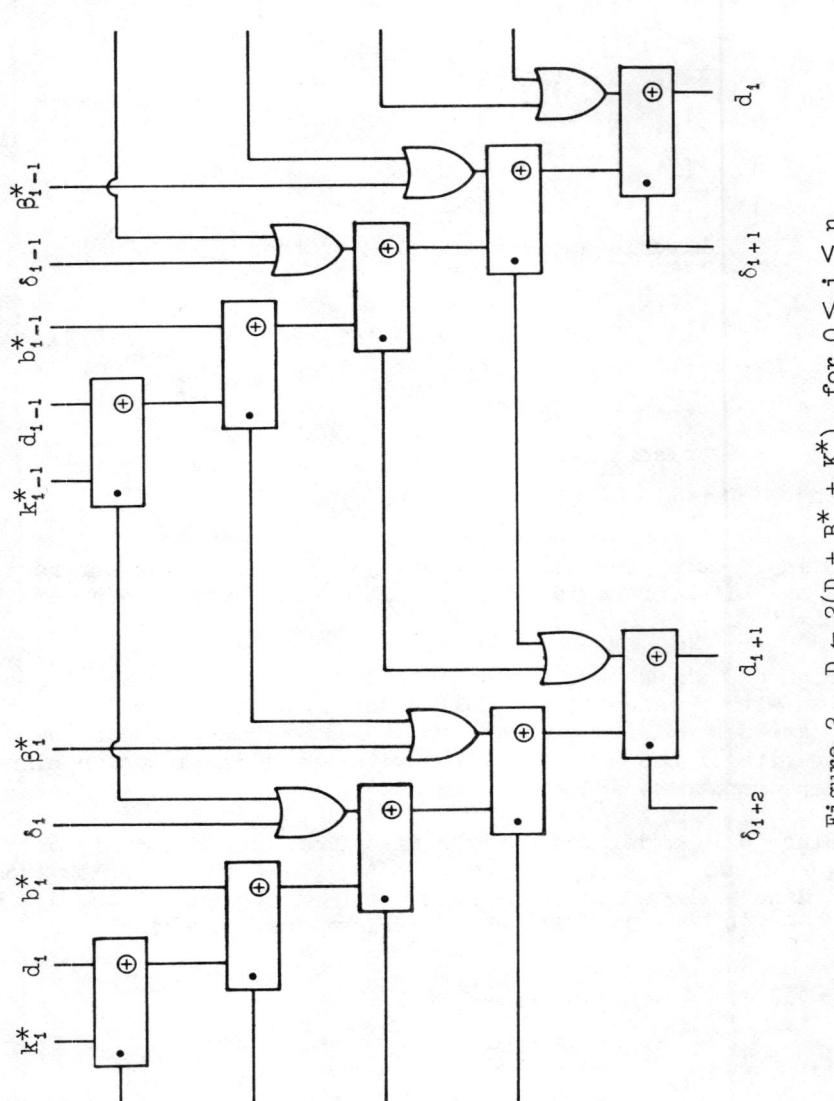

Figure 2. $D \leftarrow 2(D + B^* + K^*)$, for $0 \leq i \leq n$

A Fast Modular Multiplication Algorithm

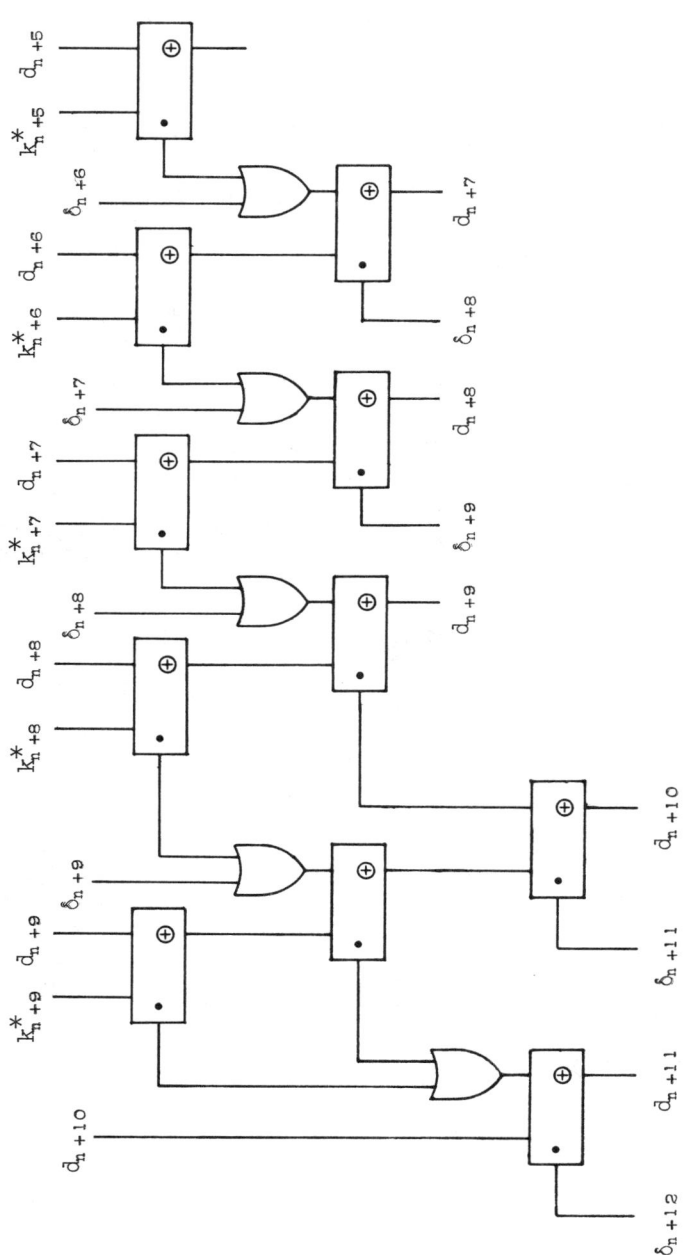

Figure 3. $D \leftarrow 2(D + B^* + K^*)$, for $i \geq n + 5$

If

(1) $$2\delta_{n+12} + d_{n+11} + \delta_{n+11} = 2t_2 + t_1 ,$$

then

(2) $$D^j \equiv 2(D^{j-1} + a_{n-1}B + \alpha_n 2B) \bmod C .$$

The remainder of the proof that Algorithm M always works consists of showing the validity of Equation (1), showing that Equation (2) can be used to prove that $D^{n+10} \equiv AB \cdot 2^{10} \bmod C$, and showing that the low-order 10 bits of D^{n+10} are all 0. Although the details will be omitted, Equation (1) is proven by first showing that $d_{n+10} d_{n+9} k_{n-2} = 0$.

Recall that $K = 2^n - C$ and $2^{n-1} < C < 2^n$. Thus $K < 2^{n-1}$ so $k_{n-1} = 0$. If $K^* = 2^{11}K$, then $k^*_{n+10} = 0$. This explains why there is no input labelled k^*_{n+10} in Figure 3. However, this algorithm can be used with any $C < 2^n$ by making the following modifications.

If $C = 2^{n-1}$, the algorithm will work as stated. Since any overflow will be a multiple of 2^{n+11}, it will be congruent to 0 mod C. Thus Equation (2) will hold.

If $C < 2^{n-1}$, let $2^{s-1} \leq C < 2^s$. Let $r = n - s$. To find AB mod C, simply find $D \equiv A \cdot (B \cdot 2^r) \bmod (2^r C)$. This is possible since $2^{n-1} \leq 2^r C < 2^n$. It can then be shown that $2^r | D$ and that $\frac{D}{2^r} \equiv AB \bmod C$.

APPLICATION

Algorithm M can be used to speed up RSA encryption. Using standard binary exponentiation [2], modular exponentiation can be performed with n squaring operations and in the worst case, n multiplies, where n is the length in bits of the modulus. If $M^{-1} \bmod C$ can be precomputed, then the number of multiplies can be reduced to n/2 in the worst case by using the following procedure.

If Hamming weight of $E \leq n/2$

    $D \leftarrow 1$
    Do $i = 1$ to n
        $D \leftarrow D^2$
        If $e_{n-i} = 1$ then $D \leftarrow D \cdot M \bmod C$
    End.

If Hamming weight of E > n/2

    $E' \leftarrow 2^n - E$

    $D \leftarrow M$

    Do   j = 1 to n

        $D \leftarrow D^2 \bmod C$

        If $e'_{n-j} = 1$ then $D \leftarrow D \cdot M^{-1} \bmod C$

    End

Table 1 is a comparison of the throughput for RSA encryption using various modular multiplication techniques. The standard technique is based on a chip designed by W. Barnard, L. Goldstein, and R. Rieden of Sandia National Laboratories, using conventional multipliers, which is capable of handling a modulus $< 2^{336}$. We assume that the clock speed for each technique is 20 MHz.

Table 1

Comparison of Throughput

	Clock Pulses	336 bit modulus		512 bit modulus	
		Chips	bits/sec	Chips	bits/sec
Standard	$\frac{n^3}{8}$	2	1,417	2	610
Simmons/Norris	$3n^2 + 2n$	3	19,802	4	13,004
Algorithm M	$2n^2 + 20n$	3	28,902	4	19,157
Algorithm M with $M^{-1} \bmod C$	$\frac{3}{2} n^2 + 15n$	3	38,535	4	25,543

REFERENCES

1. W. Diffie, and M. E. Hellman, "New directions in cryptography," *IEEE Trans. Inform. Theory*, *IT-22*, 6 (Nov. 1976), 644-654.

2. D. E. Knuth, *The Art of Computer Programming, Vol. 2: Seminumerical Algorithms*, Addison-Wesley, Reading, MA (1969).

3. R. Merkle, and M. E. Hellman, "Hiding information and receipts in trapdoor knapsacks," *IEEE Trans. Inform. Theory*, *IT-24* (Sept. 1978), 525-530.

4. M. J. Norris, and G. J. Simmons, "Algorithms for high-speed modular arithmetic," *Congressus Numeratium*, *Vol. 31* (1981), 153-163.

5. S. Pohlig, and M. E. Hellman, "An improved algorithm for computing logarithms in GF(p) and its cryptographic significance," *IEEE Trans. Inform. Theory*, *IT-24* (Jan. 1978), 106-110.

6. R. Rivest, A. Shamir, and L. Adleman, "A method for obtaining digital signatures and public-key cryptosystems," *Commun. ACM*, *21*, 2 (Feb. 1978), 120-126.

7. R. W. Roberts, National Bureau of Standards, "Encryption algorithm for computer data encryption," *Federal Register*, *Vol. 40*, No. 52 (March 17, 1975), 12134-12139.

8. J. E. Robertson, "A deterministic procedure for the design of carry-save adders and borrow-save subtractors," University of Illinois, Dept. of Computer Science, Report No. 235 (July 1967).

9. F. A. Rohatsch, "A study of transformations applicable to the development of limited carry-borrow propagation adders," Ph.D. Thesis, University of Illinois, Urbana, IL (June 1967).

10. C. S. Wallace, "A suggestion for a fast multiplier," *IEEE Trans. Elec. Comp.*, *EC-13* (Feb. 1964), 14-17.

COMPARISON OF TWO PSEUDO-RANDOM NUMBER GENERATORS [1]

Lenore Blum[2]

Dept. of Mathematics
and Computer Science
Mills College,
Oakland, Ca., and
Dept. of Mathematics
U.C. Berkeley

Manuel Blum[3]

Dept. of Electrical
Engineering and
Computer Sciences
University of California
Berkeley, CA

Michael Shub[4]

Dept. of Mathematics
Queens College
Flushing, NY
and Graduate Center
of CUNY, NY

1. Introduction

What do we want from a pseudo-random sequence generator? Ideally, we would like a pseudo-random sequence generator to quickly produce, from short seeds, long sequences (of bits) that appear in every way to be generated by successive flips of a fair coin.

Certainly, the idea of a (fast) deterministic mechanism producing such nondeterministic behavior seems contradictory: by observing its outcome over time, we could in principle eventually detect the determinism and simulate such a generator.

The resolution [Knuth], usually, is to only require of such generators that the sequences they produce pass certain standard statistical tests (e.g., in the long run, the frequency of 0's and 1's occurring in such a sequence should be nearly the same, and the 0's and 1's should be "well-mixed").

However, the usual statistical tests do not capture enough. An important property of random sequences is their unpredictability. Pseudo-random sequences should also be unpredictable to computers with feasible resources. To begin to capture this notion, we require the following definition:

DEFINITION: A pseudo-random sequence generator is *polynomial-time unpredictable* (unpredictable to the right, unpredictable to the left) or *cryptographically secure* [Shamir, Blum-Micali, Yao] if and only if the sequences it generates are not predictable (to the right, to the left) in polynomial time: i.e., given a piece of

---

[1] The final version of this paper contains the proofs of all theorems discussed here. It will appear in the SIAM Journal of Computing.

[2] This work was supported in part by the Letts-Villard Chair, Mills College.

[3] This work was supported in part by NSF grant MCS 82-04506.

[4] This work was supported in part by NSF grant MCS 82-01267.

sequence that has been produced by such a generator, but with any element (the rightmost element, the leftmost element) deleted from that piece, one can roughly do no better in guessing in polynomial time what the missing element is than by flipping a fair coin.

This notion of unpredictable is the correct one [Yao]: sequences produced by generators possessing any one of the unpredictability properties pass all polynomial time statistical tests. That is to say, these sequences cannot be distinguished by polynomial time statistical tests (with more than a negligible advantage) from sequences produced by successive flips of a fair coin.

2. Two pseudo-random sequence generators

In this paper, two pseudo-random sequence generators are defined and their properties discussed. These are called:
(1) the $1/P$ generator
(2) the $x^2 \bmod N$ generator.

The two generators are closely related. For example: **From short seeds, each quickly generates long well-distributed sequences. Both generators contain hard problems at their core** (the discrete logarithm problem and the quadratic residuosity problem, respectively). **But only the second is cryptographically secure.**

More specifically,

THEOREM 2 - Problem 4, section 6, ($1/P$): Any sequence produced by the $1/P$ generator is completely inferable; that is, given a small piece of the sequence, one can quickly infer the "seed" and efficiently extend the given piece of sequence backwards and forwards.

On the other hand,

THEOREM 4, section 7, ($x^2 \bmod N$): The $x^2 \bmod N$ generator is polynomial-time unpredictable. The sequences it generates are *provably secure:* they pass all polynomial-time statistical tests.

The $1/P$ generator has been well studied [Dickson]; its distribution properties relate to shift register sequences [Golomb]. Our observations concerning its strong inference properties, we believe, are new and surprising. The $x^2 \bmod N$ generator is new, a simplification of a generator proposed by A. Yao. Its strong security properties derive from complexity based number theoretic assumptions and arguments [Blum, Blum&Micali, Goldwasser&Micali, Yao]. Our investigation reveals additional useful properties of this generator: e.g., from knowledge of the (secret) factorization of N, one can generate the sequence backwards; from additional information about N, one can even random access the sequence. Our number-theoretic analyses also provide tools for determining the lengths of periods of the generated sequences.

Both generators have applications. The $1/P$ generator has applications to the generation of generalized de Bruijn (i.e., maximum-length shift-register) sequences. The $x^2 \bmod N$ generator has applications to public-key cryptography and to efficient computation, i.e., to converting probabilistic polynomial-time algorithms to deterministic almost-polynomial-time algorithms (see section 10, Applications).

The two generators are presented together so that each one's properties help to illuminate the other's.

3. NOTATION

Throughout this paper, $x \bmod N$ denotes the least non-negative integer remainder upon dividing x by N (rather than denoting the residue class mod N).

We use "$x^2 \bmod N$ generator" to denote a pseudo-random sequence generator, whereas "$x^2 \bmod N$" denotes the remainder upon dividing x^2 by N. A similar distinction is made between "$1/P$ generator" and the string of bits "$1/P$".

We let $|N|_b$ denote the length of N when N is expanded base b, and simply write $|N|$ when the base is clear. Note that $|N|_b = \left[1 + \log_b N\right]$.

4. THE 1/P GENERATOR

DEFINITION (the $1/P$ generator): Let P be an odd prime. Let b be a generator (i.e., primitive root) of the multiplicative group $Z_P^* = \{1, 2, ..., P-1\}$ of integers mod P.[5] The pseudo-random sequence generated by the $1/P$ *generator* with input (b, P) is the sequence of b-ary quotient digits that immediately follows the decimal point when $1/P$ is expanded to base b. We denote it by $q_1 q_2 ...$. More generally, let r_0 be any integer in the range $0 < r_0 < P$. The pseudo-random sequence generated by the $1/P$ generator with input (b, P, r_0) is the sequence of digits obtained upon dividing r_0 by P. The expansion of $1/P$ (and more generally, r_0/P) is periodic with period $P-1$: $1/P = q_1 q_2 ... q_{P-1} q_1 ...$

EXAMPLE: Let $b = 10$ and $P = 7$. This b is a primitive root mod P. The pseudo-random sequence generated by the $1/P$ generator with input (10, 7) is 142857142... since $1/7 = .142857142....$

5. THE $x^2 \bmod N$ GENERATOR

DEFINITION (the $x^2 \bmod N$ generator): Let $N = P*Q$ be a product of two distinct primes both congruent to 3 mod 4. Let x_0 be any quadratic residue in $Z_N^* = $ {integers x | 0 < x < N and gcd(x,N) = 1}. Here, N is called the *parameter* and x_0 the *seed*. The pseudo-random sequence generated by the $x^2 \bmod N$ *generator* with input (N, x_0) is the sequence of bits $b_0 b_1 \cdots$ obtained by setting $x_{i+1} = x_i^2 \bmod N$ and extracting the bit $b_i = parity(x_i)$. This sequence is periodic with period that is usually equal to $\lambda(\lambda(N))$ (see section 8 for the definition of λ and clarification of "usually"). We also note that the equality $x_i = x_0^{2^i} \bmod N = x_0^{2^i \bmod \lambda(N)} \bmod N$ enables us to efficiently compute the ith sequence element, given x_0, N and $\lambda(N)$.

EXAMPLE: Let $N=7*19=133$ and $x_0=4$. Then the sequence $x_0, x_1, ...$ has period 6; $x_0, x_1,..., x_5 = 4, 16, 123, 100, 25, 93$, and $b_0 b_1 ... b_5 = 0\ 0\ 1\ 0\ 1\ 1$. The latter string of b's is the pseudo-random sequence generated by the $x^2 \bmod N$ generator with input (133, 4). Here, $\lambda(N) = 18$ and $\lambda(\lambda(N)) = 6$.

6. THE ASSUMPTIONS

Our main results about cryptographic security follow from assumptions concerning the intractability of certain number-theoretic problems by proba-

---

[5]E. Artin's conjecture states that every integer b which is not -1 or a square is a primitive root for .3739... of all n-bit primes, asymptotically as the length n of the primes goes to infinity -- see [Shanks, p.81].

bilistic polynomial-time procedures. These results can be viewed as assertions concerning the Turing machine complexity (equivalence) of certain hard problems. Stronger results would follow from stronger assumptions concerning the circuit size complexity of the number theoretic problems below. Such results would be desirable, for example, if we wished to assure that sequences produced by our generator appear random to hard-wired circuits.

(1) THE DISCRETE LOGARITHM (INDEX FINDING) PROBLEM: Let P be a prime. Let b be a generator for Z_P^*. The function $f_{b,P} : Z_P^* \longrightarrow Z_P^*$ defined by $f_{b,P}(x) = b^x \bmod P$ is a permutation of Z_P^* that is computable in $O(|P|^3)$-time. The *discrete logarithm (index finding) problem* with parameters b and P consists in finding for each y in Z_P^* the index x in Z_P^* such that $b^x \bmod P = y$. A (probabilistic) procedure $\mathbf{P}[b, P, y]$ solves the discrete logarithm if for all primes P, for all generators b for Z_P^*, and for all y in Z_P^*, $\mathbf{P}[b,P,y] = x$ in Z_P^* such that $b^x \bmod P = y$.

THE DISCRETE LOGARITHM ASSUMPTION: (This asserts that there is a fixed fraction of time that the discrete logarithm problem cannot be solved efficiently.) Let $\mathbf{P}[b, P, y]$ be a (probabilistic) procedure for solving the discrete logarithm problem. Let $0 < \varepsilon < 1$ be a fixed constant. Let *poly* be a fixed polynomial. Then for all sufficiently large n, for all but ε-fraction of n-bit primes P, for all generators b of Z_P^*, and for at least ε-fraction of numbers $y \in Z_P^*$, $\mathbf{P}[b, P, y]$ takes more than $poly(n)$ (expected) time to output x (the particular x such that $b^x \bmod P = y$).

(2) THE QUADRATIC RESIDUOSITY PROBLEM [Gauss]: Let N be a product of two distinct odd primes. Let Z_N^* be the group of integers x, $1 \le x \le N$ and $gcd(x, N) = 1\}$, under multiplication mod N. Exactly half the elements of Z_N^* have Jacobi symbol +1, the other half have Jacobi symbol -1. Denote the former by $Z_N^*(+1)$ and the latter by $Z_N^*(-1)$. None of the elements of $Z_N^*(-1)$ and exactly half the elements of $Z_N^*(+1)$ are quadratic residues. The *quadratic residuosity problem* with parameters N and x consists in deciding, for x in $Z_N^*(+1)$, whether or not x is a quadratic residue. A probabilistic procedure $\mathbf{P}[N, x]$ solves the quadratic residuosity problem for a number N, N = a product of two distinct odd primes, and for $x \in Z_N^*(+1)$ if and only if it correctly decides whether or not x is a quadratic residue mod N (i.e., $\mathbf{P}[N, x] = 1$ if and only if x is a quadratic residue mod N).

THE QUADRATIC RESIDUOSITY ASSUMPTION: (This asserts that there is a fraction of time that the quadratic residuosity problem cannot be solved efficiently.) Let *poly*( ) be a polynomial. Let t be a positive integer. Let $\mathbf{P}[N, x]$ be any (probabilistic) *poly*-time procedure which, on inputs N, x, each of length n, outputs 0 or 1. Then for n sufficiently large and for all but $1/n^t$ fraction of numbers N of length n, N a product of two distinct odd primes, the probability that $\mathbf{P}$ decides correctly whether x is a quadratic residue mod N -- for N fixed and x selected uniformly from among all elements of $Z_N^*(+1)$ -- is less than $1 - 1/n^t$.

7. THE 1/P GENERATOR IS INFERABLE

Let P and b be relatively prime positive integers and r_0 an integer in the range $0 < r_0 < P$. Denote the expansion of r_0/P to base b by

$$r_0/P = .q_1 q_2 q_3 \cdots \qquad (1)$$

where $0 \le q_i < b$. Since b is prime to P, the expansion is periodic. Then, for

$m \geq 0$,
$$(b^m \cdot r_0)/P = q_1 \cdots q_m \cdot q_{m+1} q_{m+2} \cdots = (q_1 \cdots q_m) + r_m/P \qquad (2)$$
where
$$0 < r_m = b^m r_0 \bmod P < P \qquad (3)$$
and
$$0 < r_m/P = .q_{m+1} q_{m+2} \cdots = (b^m \cdot r_0/P) \bmod 1 < 1 \qquad (4)$$

Here, $q_1, q_2, \cdots$ are (quotient) *digits* base b and $q_1 q_2 \cdots$ denotes their concatenation, whereas r_m, the mth remainder (of r_0/P base b), is an *integer* whose length (base b) is less than or equal to the length of P: $|r_m| \leq |P|$. Recall from the definition of the $1/P$ generator in section 3 that, for P prime and b a primitive root mod P, eq. 1 defines the pseudo-random sequence generated by the $1/P$ generator with input (b, P, r_0).

There are several reasons one might consider the $1/P$ generator a good pseudo-random sequence generator: the sequences produced have long periods and nice distribution properties (Theorem 1 below). In addition, these sequences possess certain hard-to-infer properties. For example, given a remainder r generated during the expansion of $1/P$ base b, it is hard, in general, to find any index m such that $r_m = r$. This is because $r_m = b^m \bmod P$, so m is the discrete logarithm of r mod P. It follows (Theorem 2, problem 1) that, given a string of quotient digits $q_{m+1} q_{m+2} \cdots q_{m+k}$ ($k \leq poly(|P|)$), it is hard in general to find its location in the sequence. Thus, these strings appear to be "well mixed."

On the other hand, Theorem 2 will give a sense, which is correct, that the $1/P$ generator yields a poor pseudo-random sequence: from knowledge of P and any $|P|$-long segment of sequence, one can efficiently extend the segment backwards and forwards (problem 2). More surprisingly (problem 4), from knowledge of any $2|P|+1$ successive elements of the sequence, but *not* P, one can efficiently reconstruct P, and hence efficiently continue the sequence in either direction.

It follows that there is a simple efficient statistical test for deciding whether a 3n-long string of digits has either been extracted from $1/P$, for some prime P of length n, or has been generated at random (uniform probability distribution), given that it was produced in one of those two ways: Use 2n+1 of the given 3n digits to recover the suspected P; use this P to generate 3n digits; then compare the generated digits with the 3n given digits: if they agree, the string has probably (with probability $\geq 1 - 1/2^{n-1}$) been generated using the $1/P$ generator.

To lead up to Theorem 1, we consider the following type of sequences (closely related to maximum-length shift register sequences [Golomb]).

DEFINITION: Let P, b denote arbitrary positive integers. A *(generalized) de Bruijn sequence of period P-1, base b*, is a sequence $q_1 q_2 \cdots$ of b-ary digits (i.e., $0 \leq q_i < b$ for all i) of period P-1 such that
(1) every b-ary string of length $|P|$-1 occurs at least once in the sequence, and
(2) every b-ary string of length $|P|$ occurs at most once in any given period of the sequence.

THEOREM 1
  Let P = prime. Let $b \in \{1,2,...,P-1\}$ be a primitive root mod P. Let $r_0 \in \{1,2,...,P-1\}$. Then the pseudo-random sequence generated by the $1/P$ generator with input (b, P, r_0) is a (generalized) de Bruijn sequence of period P-1, base b.

PROOF:
Since $r_m = b^m \cdot r_0 \bmod P$ and b is a primitive root mod P, the sequence of remainders r_m (generated during the expansion of $1/P$) is periodic with period $P-1$, the remainders in any period are distinct, and $\{r_m \mid 1 \leq m \leq P-1\} = \{1, 2, \ldots, P-1\}$.

Similarly, the sequence of quotients r_m/P is periodic with period $P-1$, the quotients in any period are distinct, and

$$\{r_m/P \mid 1 \leq m \leq P-1\} = \{1/P, 2/P, \ldots, (P-1)/P\}. \tag{5}$$

Therefore, the sequence of quotient digits q_m is periodic with period at most $P-1$. If the period were less than $P-1$, then there would be integers $0 \leq m_1 < m_2 < P-1$ such that $\cdot q_{m_1+1} q_{m_1+2} \cdots = \cdot q_{m_2+1} q_{m_2+2} \cdots$. Since $r_m/P = \cdot q_{m+1} q_{m+2} \cdots$, we would have $r_{m_1}/P = r_{m_2}/P$, a contradiction. Therefore the period is $P-1$. [Gauss]

Now, a string $a_1 \cdots a_s$ of s b-ary digits appears somewhere in the expansion of r_0/P if and only if it appears as an initial string in the expansion of r_m/P for some $1 \leq m \leq P-1$ if and only if (by eq. 5) it appears as an initial string in the expansion of k/P for some $1 \leq k \leq P-1$. But also, the set of b-ary strings of length s correspond exactly to the subintervals of the unit interval $[0,1)$ of the form $[l/b^s, (l+1)/b^s)$ where l is an integer, $0 \leq l < b^s$. Since $1/P < 1/b^{|P|-1}$, there is for each l, at least one k, $1 \leq k \leq P-1$ such that $k/P \in [l/b^{|P|-1}, (l+1)/b^{|P|-1})$ and so we have property 1. Since $1/b^{|P|} < 1/P$, there is for each l at most one k, $1 \leq k \leq P-1$ such that $k/P \in [l/b^{|P|}, (l+1)/b^{|P|})$, and so we get property 2.
QED

So, if P is prime and b is a primitive root mod P, it follows from Theorem 1 concerning de Bruijn property 1 (and Artin's conjecture -- see footnote concerning that conjecture). that neither $|P|-1$ successive digits of quotient, $q_{m+1} \cdots q_{m+|P|-1}$, nor (the approximately $|P|-1$ successive digits of) a remainder, r_m, are enough to construct P, or to extend the sequence, on purely information-theoretic grounds. In contrast, it will follow from Theorem 2 below that (various combinations of) approximately $2|P|$ digits of information are sufficient to efficiently extend the sequence in either direction.

THEOREM 2
Let P and b be relatively prime integers > 1 (P not necessarily prime!), and let r_0 be an integer in the range $0 < r_0 < P$. The following problems are solvable in polynomial($|P|$)-time:

PROBLEM 1
Choose a polynomial, $poly(\ )$, and hold it fixed.
INPUT: P, b, remainder r_m, positive integer $k \leq poly(|P|)$.
OUTPUT: r_{m-1}, r_{m+k}; $q_m q_{m+1} \cdots q_{m+k}$.

PROBLEM 2 [Gauss]
This is a computational version of Theorem 1 concerning De Bruijn property 2. (A similar algorithm gives the computational version of property 1.)
INPUT: P, b, $|P|$ successive digits of quotient $q_{m+1} q_{m+2} \cdots q_{m+|P|}$.
OUTPUT: r_m (and hence, by problem 1, $r_{m+|P|}$ and q_m, $q_{m+|P|+1}$).

PROBLEM 3
We assume that P is relatively prime to each of $1,2,\ldots,b$ (to ensure that the output is the unique P that generated r_m and r_{m+1}).
INPUT: b, r_m, r_{m+1} such that $r_m \cdot b \neq r_{m+1}$ (i.e. $r_m \geq P/b$).
OUTPUT: P (and therefore also, by problem 1, $q_m q_{m+1} \cdots q_{m+|P|}$).

PROBLEM 4
We assume that r_0 is relatively prime to P (e.g., $r_0 = 1$).
INPUT: b; k quotient digits, $q_{m+1} q_{m+2} \cdots q_{m+k}$, where $k = \lceil \log_b(2P^2) \rceil$ and m is arbitrary. (Note that $k \leq 2|P|+1$)[6].
OUTPUT: P; r_m (and hence by Problem 1, q_m and q_{m+k+1}).

PROOF:

To solve problem 1: $r_{m+k} = b^k r_m \bmod P$ and $r_{m-1} = b^{-1} r_m \bmod P$ where b^{-1} is the inverse of $b \bmod P$. We note that

$$(b^k r_m)/P = q_{m+1} \cdots q_{m+k} + r_{m+k}/P \tag{6}$$

So, $q_m \cdots q_{m+k} = \lfloor (b^{k+1} r_{m-1})/P \rfloor$. (By convention, we do not drop initial digits in a concatenation of quotient digits, e.g., in eq. 6.)

To solve problem 2: By eq. 6, $r_m = \dfrac{(q_{m+1} \cdots q_{m+|P|}) \cdot P}{b^{|P|}} + \dfrac{r_{m+|P|}}{b^{|P|}}$. Since $r_{m+|P|} < P < b^{|P|}$, $r_m = \left\lceil \dfrac{(q_{m+1} \cdots q_{m+|P|}) \cdot P}{b^{|P|}} \right\rceil$.

In problems 3 and 4, the number P is not available and must be constructed.

To solve problem 3: By eq. 6 with $k = 1$, $b \cdot r_m - r_{m+1} = q_{m+1} \cdot P$ where $0 \leq q_{m+1} < b$. Actually, $0 < q_{m+1}$, since, by assumption, $b \cdot r_m \neq r_{m+1}$. Therefore, P equals some integer in the sequence of real numbers $\dfrac{b \cdot r_m - r_{m+1}}{1}$, $\dfrac{b \cdot r_m - r_{m+1}}{2}, \ldots, \dfrac{b \cdot r_m - r_{m+1}}{b-1}$. Select any integer P in the sequence such that P is relatively prime to $1,2,\ldots,b$. Such an integer P is unique; for suppose to the contrary that P, Q are two such integers relatively prime to each of $1,2,\ldots,b$. Then $P \cdot (i) = Q \cdot (j)$ for some $0 < i,j < b$. Without loss of generality, suppose $P < Q$. Q is relatively prime to each of $1, 2, \ldots, b$, so $\gcd(Q, i) = 1$, so $Q | P$, so $Q \leq P$, which is a contradiction.

The solution to problem 4, which is very pretty, is by continued fractions:
By eq. 6, $\dfrac{r_m}{P} = \dfrac{q_{m+1} \cdots q_{m+k}}{b^k} + \varepsilon$ where $0 \leq \varepsilon < \dfrac{1}{b^k}$.
By LeVeque p. 237 Theorem 9.10, the continued fraction expansion of $\dfrac{q_{m+1} \cdots q_{m+k}}{b^k}$ has convergent $\dfrac{r_m}{P}$ if $\dfrac{1}{b^k} \leq \dfrac{1}{2P^2}$, i.e., $2P^2 \leq b^k$, i.e., $\log_b(2P^2) \leq k$, as postulated. Since both b and r_0 are relatively prime to P, it follows (from eq. 3) that $\gcd(r_m, P) = 1$, so $r_m = A_i$ and $P = B_i$ So $\dfrac{r_m}{P} = \dfrac{A_i}{B_i}$ for one of the convergents $\dfrac{A_1}{B_1}$, $\dfrac{A_2}{B_2}, \ldots$ of the fraction $\dfrac{q_{m+1} \cdots q_{m+k}}{b^k}$. Since both

---

[6] Much as one wants, this result cannot be improved to permit $k = 2|P|$. For example, for $b = 10$ and $P = 97$ ($1/97 = .010309\ldots$), the four digits 1030 do not yield P.

b and r_0 are relatively prime to P, it follows (from eq. 3) that $gcd(r_m, P) = 1$, so $r_m = A_i$ and $P = B_i$.

It remains to show that r_m and P can be obtained by generating the above convergents until the first k digits of $\dfrac{A_i}{B_i}$ are $q_{m+1} \cdots q_{m+k}$, at which point $r_m = A_i$ and $P = B_i$. To see why, recall that the continued fraction $\dfrac{q_{m+1} \cdots q_{m+k}}{b^k} = 1/a_1 + 1/a_2 + 1/a_3 + \ldots 1/a_i + \ldots$ has convergents $\dfrac{A_1}{B_1} = \dfrac{1}{a_1}$, $\dfrac{A_2}{B_2} = \dfrac{a_2}{a_1 a_2 + 1}, \ldots, \dfrac{A_i}{B_i} = \dfrac{a_i A_{i-1} + A_{i-2}}{a_i B_{i-1} + B_{i-2}}, \ldots$ Here, the B_i are strictly increasing with i. Since for some i, $A_i / B_i = r_m / P$, this procedure for obtaining r_m and P will never go beyond $A_i / B_i = r_m / P$. To see that the procedure generates convergents to the point where $A_i / B_i = r_m / P$, note that when $A_j / B_j = \cdot q_{m+1} \cdots q_{m+k} \cdots$, the error is sufficiently small to ensure that $A_j / B_j = r_m / P$.

Since A_i and B_i grow exponentially, $P = B_i$ and $r_m = A_i$ can be computed in polynomial($|B_i|$), in particular in O(number of steps to compute the ith Fibonacci number), and therefore in polynomial($|P|$) steps. This solves problem 4.
QED

REMARK: The solution to problem 4 can be viewed as a computational version of the following: for positive integers $b > 1$, k and l, $0 \le l < b^k$, there is at most one integer $P > 1$ with $gcd(b, P) = 1$ and $2P^2 \le b^k$, and at most one integer r with $1 \le r \le P-1$ and $gcd(r, P) = 1$, such that $r/P \in [l/b^k, (l+1)/b^k)$.

EXAMPLE: Let $b = 10$ and $P = 503$. Then P is prime and b is a primitive root mod P, so the $1/P$ generator with input $(10, 503)$ quickly generates a sequence of base 10 digits with period 502. This sequence is
00198 80715 70576 54075 54671 96819 08548 70775 34791 25248 50894 63220 67594 **43339** 96023 85685 88469 18489 06560 63618 29025 84493 04174 95029 82107 35586 48111 33200 79522 86282 30616 30218 68787 27634 19483 10139 16500 99403 57852 88270 37773 35984 09542 74353 87673 95626 24254 47316 10337 97216 69980 11928 42942 34592 44532 80318 09145 12922 46520 87475 14910 53677 93240 55666 00397 61431 41153 08151 09343 93638 17097 41550 69582 50497 01789 26441 35188 86679 92047 71371 76938 36978 13121 27236 58051 68986 08349 90059 64214 71172 96222 66401 59045 72564 61232 60437 37574 55268 38966 20278 33001 98807 ...

Since $|503| = 3$, every string of two decimal digits occurs at least once in the above sequence, and every string of three decimal digits occurs at most once in any period of the sequence.

Since $k = \lceil \log_{10}(2 \cdot 503^2) \rceil = 6$, we can, from any segment of length 6 of the the above sequence, efficiently recover P, and then quickly extend the segment in either direction. For example, consider the segment 433399 (shown in bold type above). The continued fraction expansion of $433,399/1,000,000$ is $433,399/1,000,000 = 1/2 + 1/3 + 1/3 + 1/1 + 1/16 + 1/6 + 1/1 + 1/1 + 1/358 + \ldots$, and its first five convergents are: $1/2 = .5$; $3/7 = .48\ldots$; $10/23 = .434\ldots$; $13/30 = .43333\ldots$; $218/503 = .4333999\ldots$. At last, the first 6 digits agree with the segment 433399. So we get $P = 503$ and $r_m = 218$ (and so $r_{m-1} = 10^{-1} \cdot r_m \bmod 503 = 151*218 \bmod 503 = 223$). In this way, we can extend the given segment, 433399, forwards and backwards.

6. THE $x^2 \bmod N$ GENERATOR IS UNPREDICTABLE

In this section we elaborate on properties of the $x^2 \bmod N$ pseudo-random sequence generator, and prove that it is polynomial-time unpredictable (Theorem 4, this section).

First we recall some number-theoretic facts. Suppose $N = P \cdot Q$ where P and Q are distinct odd primes. Let $Z_N^* = \{$integers $x \mid 0 < x < N$ and $gcd(x, N) = 1\}$. Then QR_N, the set of quadratic residues mod N, form a multiplicative subgroup of Z_N^* of order $\varphi(N)/4$ (where $\varphi(N)$ is the cardinality of Z_N^*). Each quadratic residue $x^2 \bmod N$ has four distinct square roots, $\pm x \bmod N$, $\pm y \bmod N$. If we also assume, as we shall for the rest of this paper, that $P \equiv Q \equiv 3 \bmod 4$, then each quadratic residue mod N has *exactly* one square root which is also a quadratic residue (see Lemma 1, this section). In other words, squaring mod N is a 1-1 map of QR_N onto QR_N. (Comment: half the primes of length n are congruent to 3 mod 4 asymptotically as $n \to \infty$ [LeVeque], so there are plenty such N.)

We now investigate what properties can be inferred about sequences produced by the $x^2 \bmod N$ generator, given varying amounts of information. In the following, N is of the *prescribed form*, that is to say, $N = P*Q$ where P, Q are distinct primes both congruent to 3 mod 4. Also, x_i is a quadratic residue mod N, $x_{i+1} = x_i^2 \bmod N$ and $b_i = parity(x_i)$:

1. Clearly, knowledge of N is sufficient to efficiently generate sequences $x_0, x_1, x_2, \cdots$ (and hence sequences $b_0 b_1 b_2 \cdots$) in the forward direction, starting from any given seed x_0. The number of steps per output is $O(|N|^{1+\epsilon})$ using fast multiplication.

    Conversely, it follows from more general results of [Plumstead], that there is a polynomial *poly* such that from knowledge of n, and any sequence $x_0, \ldots, x_k$, $k = poly(n)$, generated by x_0 and an unknown N of length n, we can infer in $poly(N)$ - time an N (of length n) that produces this sequence.

2. Given N, the factors of N are necessary and sufficient to efficiently generate the $x^2 \bmod N$ sequences in the reverse direction, $x_0, x_{-1}, x_{-2}, \cdots$, starting from any given seed x_0. (See proof below)

3. What is more, the factors of N are necessary -- assuming they are necessary for deciding quadratic residuosity of an x in $Z_n^*(+1)$ -- to have even an ε-advantage in guessing in polynomial time the parity of x_{-1}, given N and given x_0 chosen "at random" from QR_N. (Note that to choose a quadratic residue at random with the uniform probability distribution from QR_N, it is sufficient to choose x at random (with the uniform probability distribution) from Z_N^* and square it mod N).[7]

To see Claim 2 above, we first prove the following

LEMMA 1

If $N = P \cdot Q$ where P and Q are distinct primes such that $P \equiv Q \equiv 3 \bmod 4$, then each quadratic residue mod N has exactly one square root which is a quadratic residue.

PROOF:
Whenever N is a product of two distinct odd primes, every quadratic residue

---

[7] A more formal statement of claim 3 will appear in the final version of this paper.

mod N has four square roots, $\pm x$ and $\pm y$. Since $N \equiv 1 \bmod 4$, their Jacobi symbols satisfy $(\frac{+x}{N}) = (\frac{-x}{N})$ and $(\frac{+y}{N}) = (\frac{-y}{N})$. Since $P \equiv 3 \bmod 4$, $(\frac{+x}{N}) \neq (\frac{+y}{N})$ (this can easily be proved from the fact that $gcd(x+y, N) = P$ and $gcd(x-y, N) = Q$). Thus $(\frac{+x}{N}) = (\frac{-x}{N}) \neq (\frac{+y}{N}) = (\frac{-y}{N})$. Eliminating the two roots, say $\pm y$, with Jacobi symbol -1 with respect to N, we are left with the two roots $\pm x$ having Jacobi symbol +1 with respect to N. Exactly one of these roots has Jacobi symbol +1 with respect to both P and Q, because $P \equiv 3 \bmod 4$, and this one and this one only is a quadratic residue mod N.
QED

The necessity (of knowing the factors of N) now follows: Suppose we can efficiently generate such sequences in the reverse direction. To factor N, select an x in Z_N^* whose Jacobi symbol is $(\frac{x}{N}) = -1$. Let $x_0 = x^2 \bmod N$ and compute x_{-1}. Then efficiently compute $gcd(x+x_{-1}, N) = P$ or Q. We can sharpen this argument to show [Rabin] that the ability to compute x_{-1} for even a fraction of seeds x_0 will enable us to factor N efficiently with high probability.

On the other hand, if we know the factors of N we can use the algorithm described in Theorem 3 (below) to efficiently generate sequences backwards:

THEOREM 3
There is an efficient deterministic algorithm A which when given N (of the prescribed form), the prime factors of N and any quadratic residue x_0 in Z_N^*, efficiently computes the unique quadratic residue x_{-1} mod N such that $(x_{-1})^2 \bmod N = x_0$. Thus,
$A(P,Q,x_0) = x_{-1}$.

PROOF:
By Lemma 1, the map from the quadratic residues mod N into the quadratic residues mod N, $f: x \to x^2 \bmod N$, is 1-1 onto.
The algorithm A can now be described as follows:
INPUT: P, Q = two distinct primes congruent to 3 mod 4; x_0 = a quadratic residue mod N, where $N = P \cdot Q$.
OUTPUT: A quadratic residue x_{-1} mod N whose square mod N is x_0.
Compute $x_P = \sqrt{x_0} \bmod P$ such that $(\frac{x_P}{P}) = +1$, where $\sqrt{x_0} \bmod P$ denotes an integer in Z_P^* whose square mod P is x_0 (this computation of $x_P = \sqrt{x_0} \bmod P$ can be done efficiently by a deterministic polynomial-time algorithm). Compute $x_Q = \sqrt{x_0} \bmod Q$ such that $(\frac{x_Q}{Q}) = +1$. Use the Euclidean algorithm to construct integers u, v such that $P \cdot u + Q \cdot v = 1$, and from that obtain the particular number, $x_N = \pm x_P \cdot Q \cdot v \pm x_Q \cdot P \cdot u = \sqrt{x_0} \bmod N$, that is a square root of $x_0 \bmod N$, and that is also a quadratic residue with respect to both P and Q and therefore with respect to N.
QED

To see Claim 3 above, we start with the following

DEFINITION: Given a polynomial $poly(\ )$ and $0 < \varepsilon \leq 1/2$, a 0-1 valued probabilistic $poly$-time procedure $P(\ ,\ )$ has an ε-*advantage for N in guessing (determining) parity* (of x_{-1} given arbitrary x_0 in QR_N) if and only if given x_0 selected uniformly from QR_N, $Prob[\ P(N, x_0) = Parity(x_{-1})] \geq 1/2 + \varepsilon$.

In a similar fashion, we can define a procedure having an ε-*advantage for N in guessing quadratic residuosity* (of arbitrary $x \in Z_N^*(+1)$) [Goldwasser-Micali]. In this regard, the $1/2 + \varepsilon$ makes sense since exactly half the elements in $Z_N^*(+1)$ are quadratic residues.

LEMMA 2
   An ε-advantage for determining parity (of x_{-1} given quadratic residue x_0) can be converted, efficiently and uniformly, to an ε-advantage for determining quadratic residuosity (of x in $Z_N^*(+1)$).[8]

PROOF
Let $x \in Z_N^*(+1)$ be an element whose quadratic residuosity mod N is to be determined. Set $x_0 = x^2 \bmod N$. Since $P \equiv Q \equiv 3 \bmod 4$, the square roots of $x^2 \bmod N$ that are in $Z_N^*(+1)$ are $\pm x$ (see proof of Lemma 1), and since N is odd, each of these square roots has opposite parity. Only one of these square roots is a quadratic residue (i.e., equal to x_{-1}), and only one of these has parity equal to $parity(x_{-1})$. Therefore, x is a quadratic residue mod N if and only if $x = x_{-1}$ if and only if $parity(x) = parity(x_{-1})$.
QED

LEMMA 3 (Goldwasser and Micali)
   An ε-advantage for determining quadratic residuosity can be amplified as much as we like, uniformly and efficiently.[9]

IDEA OF PROOF
Let $x \in Z_N^*(+1)$ be an element whose quadratic residuosity mod N is to be determined. To this end, select r at random with uniform probability from Z_N^*. Compute $x \cdot r^2 \bmod N$. [Comment: For $x \in QR_N$, $x \cdot r^2 \bmod N$ is uniformly distributed over QR_N; for $x \notin QR_N$, $x \cdot r^2 \bmod N$ is uniformly distributed over $Z_N^*(+1) - QR_N$.] Test each of the resulting numbers, $x \cdot r^2 \bmod N$, for quadratic residuosity. Taking the majority vote amplifies the advantage as much as one likes.
QED

Claim 3 follows: Suppose to the contrary that **P** is a probabilistic *poly* procedure that has a $1/n^t$ advantage in determining parity (for infinitely many n, and for more than $1/n^t$ of prescribed numbers N of length n). Then convert **P** (Lemma 2) to a probabilistic *poly'* procedure **P'** for determining quadratic residuosity that has an amplified advantage (Lemma 3) of $1/2 - 1/n^{t'}$ (for these same integers N). This contradicts the quadratic residuosity assumption.

   Leading up to Theorem 4 we make the following

DEFINITION:
1. A *predictor* **P** (.) for the $x^2 \bmod N$ generator is a probabilistic *poly*-time procedure that on inputs N, $b_1 \cdots b_k$, with $b_i \in \{0,1\}$ and $k \le poly(|N|)$, outputs a 0 or 1.

2. **P** has an ε-*advantage* for N in predicting (to the left) sequences produced by the $x^2 \bmod N$ generator if and only if $Prob\ [\ \mathbf{P}(N, b_1 \cdots b_k) = b_0\ |$ seed x_0 is selected uniformly from QR_N, $k \le poly(|N|)$, and $b_1 \cdots b_k$ is the

---

[8] A more formal statement of Lemma 2 will appear in the final version of this paper.

[9] A more formal statement of Lemma 3 will appear in the final version of this paper.

segment of the sequence generated by the $x^2 \bmod N$ generator with input (N, x_0)] $\geq 1/2 + \varepsilon$.

THEOREM 4

The $x^2 \bmod N$ generator is an unpredictable (cryptographically secure) pseudo-random sequence generator. That is to say, for each probabilistic poly-time predictor **P**, and positive integer t, **P** has at most a $1/n^t$ advantage for N in predicting sequences to the left (for sufficiently large n and for all but $1/n^t$ prescribed numbers N of length n).

PROOF:
Suppose we have a predictor for the $x^2 \bmod N$ generator with an ε - advantage for N. This can be converted efficiently and uniformly into a procedure with an ε-advantage in guessing parity (of x_{-1} given arbitrary x_0 in QR_N). To see this, suppose we are given $x_0 \in QR_N$. From seed x_0 generate the sequences $b_0 b_1 b_2 \cdots$. Then $parity(x_{-1}) = b_{-1}$.

Now convert (Lemma 2) to a procedure for guessing quadratic residuosity with an amplified advantage (Lemma 3) to get a contradiction to the Quadratic Residuosity Assumption.
QED

It follows from Yao's theorem that the sequences produced by the $x^2 \bmod N$ generator pass every probabilistic polynomial-time statistical test. Yao's theorem says, in essence, that the unpredictability property is a universal test for randomness. The *idea* of his argument is as follows. Suppose there were a probabilistic poly-time test T that has an advantage in distinguishing between the pseudo-random sequences produced by an unpredictable generator and truly random sequences of bits. Then, given $k \leq poly(n)$, we can find j (in probabilistic $poly(n)$-time) such that T has an advantage in distinguishing between sequences in $A = \{r_1 \cdots r_j b_0 \cdots b_k\}$ and $B = \{r_1 \cdots r_j r_{j+1} b_1 \cdots b_k\}$, where the $b_0 \cdots b_k$ are sequences produced by the generator, and the $r_1 \cdots r_{j+1}$ are sequences of independent random bits.

We can convert T into a predictor for the generator: Given a sequence $b_1 \cdots b_k$ produced by the generator, we pass a $poly(n)$ sample of sequences of the form $r_1 \cdots r_j 0 b_1 \cdots b_k$ (where the $r_1 \cdots r_j$ are random) through test T. If $b_0 = 0$, then T is more likely to say these sequences belong to A, in which case we predict 0 for b_0. If $b_0 \neq 0$, then the initial segments of these sequences are weighted more heavily on the random side, and thus T is more likely to say they belong to B, in which case we predict 1 for b_0. T's advantage in distinguishing between pseudo-random and random sequences is thus converted into an advantage in predicting b_0 correctly.

REMARK: We can construct another unpredictable generator as follows: recall that since $N \equiv 1 \bmod 4$, both x and $-x$ (in Z_N^*) have the same Jacobi symbol, and since N is odd, x and $-x$ have opposite parity. Therefore, the parity property partitions $Z_N^*(+1)$ in half. In a similar fashion, the location property, where $location(x) = 0$ if $x < (N-1)/2$, 1 if $x \geq (N-1)/2$, partitions $Z_N^*(+1)$ in half. Thus we get the following

THEOREM: The modified $x^2 \bmod N$ generator, gotten by extracting the location bit at each stage (instead of parity) is cryptographically secure (modulo the Quadratic Residuosity assumption).

CONJECTURE: The modified $x^2 \bmod N$ generator, gotten by extracting two bits at each stage, $parity(x)$ and $location(x)$, is cryptographically secure.

QUESTION: *Parity*(x) is the least significant bit of x; we can think of *location*(x), in a sense, as the most significant bit. How many bits (and which ones) can we extract at each stage and still maintain cryptographic security?

9. LENGTHS OF PERIODS
(OF THE SEQUENCES PRODUCED BY THE $x^2 \bmod N$ GENERATOR)

What exactly is the period of the sequence generated by the $x^2 \bmod N$ generator? The question arises as soon as one starts to construct examples. Let $\pi(x_0)$ be the period of the sequence $x_0, x_1, x_2, \cdots$. Since the $x^2 \bmod N$ generator is an unpredictable pseudo-random sequence generator, it follows that on the average, $\pi(x_0)$ will be long. In this section we investigate the precise lengths of these periods. To start, we show that the period is a divisor of $\lambda(\lambda(N))$.

DEFINITION: Let $M = 2^e * P_1^{e_1} * \cdots * P_k^{e_k}$, where $P_1, \ldots, P_k$ are distinct odd primes.
Carmichael's λ-function is defined by $\lambda(2^e) = \begin{cases} 2^{e-1} \text{ if } e = 1 \text{ or } 2 \\ 2^{e-2} \text{ if } e > 2 \end{cases}$ and
$\lambda(M) = \text{lcm}[\lambda(2^e), (P_1-1)*P_1^{e_1-1}, \ldots, (P_k-1)*P_k^{e_k-1}]$.
Carmichael [LeVeque, Knuth] proves that $\lambda(M)$ is both the least common multiple and the supremum of the orders of the elements in Z_M^*.

The following theorem asserts that the period, $\pi(x_0)$, divides $\lambda(\lambda(N))$.

THEOREM 5:
Let N be a number of the prescribed form (that is to say, $N = P*Q$ where P, Q are distinct primes both congruent to 3 mod 4). Let x_0 be a quadratic residue mod N. Let $\pi = \pi(x_0)$ = period of the sequence $x_0, x_1, x_2, \cdots$. Then $\pi | \lambda(\lambda(N))$.

PROOF: To appear in the final version of this paper.

The following theorem provides conditions under which $\lambda(\lambda(N)) | \pi(x_0)$ -- and therefore $\lambda(\lambda(N)) = \pi(x_0)$.

THEOREM 6
Let N be a number of the prescribed form, x_0 a quadratic residue mod N, $\pi(x_0)$ = period of the sequence $x_0, x_1, \cdots$.
1. Choose N so that $\text{ord}_{\lambda(N)/2}(2) = \lambda(\lambda(N))$. (Note: this equality frequently holds for prescribed N. See below and Theorem 7.)
2. Choose quadratic residue x_0 so that $\text{ord}_N(x_0) = \lambda(N)/2$. (Note: one can always choose a quadratic residue x_0 this way. See below.)
Then $\lambda(\lambda(N)) | \pi(x_0)$ (and therefore $\lambda(\lambda(N)) = \pi(x_0)$).

PROOF: To appear in the final version of this paper.

Condition 2 of the above theorem holds for a substantial fraction of quadratic residues, x_0 in Z_N^*. Specifically, the number of quadratic residues in Z_N^* that are of order $\frac{\lambda(N)}{2} \bmod N$ is $\Omega\left[\frac{N}{(\ln\ln N)^2}\right]$ (where $f(n) = \Omega(g(n))$ means there exists a constant c such that $f(n) > c \cdot g(n)$ for all sufficiently large n). To derive this lower bound, let $N = P \cdot Q$. Let g_P, g_Q be generators mod P, Q respectively. Let $a \equiv g_P \bmod P$, $\equiv g_Q \bmod Q$. It is easy to see that $\text{ord}_N a =$

$lcm[P-1, Q-1] = \lambda(N)$. Now there are $\varphi(\varphi(P))$ generators mod P and $\varphi(\varphi(Q))$ generators mod Q. By the Chinese Remainder Theorem, $Z_N^* = Z_P^* \times Z_Q^*$, so there are at least $\varphi(\varphi(P)) * \varphi(\varphi(Q))$ elements in Z_N^* of order $\lambda(N)$. But $\varphi(x) > \frac{x}{6 \ln \ln x}$ for all integers $x > 2$. Hence $\varphi(\varphi(P)) * \varphi(\varphi(Q)) = \varphi(P-1) * \varphi(Q-1) \geq \frac{P-1}{6 \ln \ln(P-1)} \cdot \frac{Q-1}{6 \ln \ln(Q-1)} \geq \frac{N-P-Q+1}{[6 \ln \ln(N-1)]^2} = \Omega \left[\frac{N}{(\ln \ln N)^2} \right]$. The map $x \to x^2 \bmod N$ is 4:1. Therefore, there are at least $\Omega \left[\frac{N}{4(\ln \ln N)^2} \right]$ quadratic residues in Z_N^* of order $\lambda(N)/2$.

Condition 1 of the above theorem is harder to ensure in general. The following definition and theorem give conditions of special interest for our applications, under which condition 1 will hold.

DEFINITION: A prime P is *special* if $P = 2P_1 + 1$ and $P_1 = 2P_2 + 1$ where P_1, P_2 are odd primes. A number $N = P*Q$ is a *special* number of the prescribed form if and only if P, Q are distinct odd primes both congruent to 3 mod 4, and P, Q are both special (note: distinctness of P and Q implies that $P_2 \neq Q_2$).

EXAMPLE: The primes 2879, 1439, 719, 359, 179, 89, are special. The number $N = 23*47$ is a special number of the prescribed form.

REMARK: It is reasonable to expect [cf. Shanks] that the fraction of n-bit numbers that are special primes is asymptotically $1/((\ln P)(\ln P_1)(\ln P_2))$ which is asymptotically $1/(n^3 \ln^3 2)$, since $2^n < P < 2^{n+1}$, $2^{n-1} < P < 2^n$, and $2^{n-2} < P < 2^{n-1}$. It follows that there is an efficient, i.e., *polynomial*(n), probabilistic algorithm to find special n-bit primes: simply generate n bit numbers at random and use the Miller-Rabin probabilistic primality test [Miller, Rabin] to select the ones that are special.

THEOREM 7
Suppose N is a special number of the prescribed form, and that 2 is a quadratic residue with respect to at most *one* of P_1, Q_1.[10] Then $ord_{\lambda(N)/2}(2) = \lambda(\lambda(N))$ (and therefore $\lambda(\lambda(N)) = \pi(x_0)$ for some x_0).

PROOF: To appear in the final version of this paper.

OPEN QUESTION: Let $\pi_b(x_0)$ be the period of the sequence $b_0 b_1 \cdots$ produced by the $x^2 \bmod N$ generator with input (N, x_0). Then $\pi_b(x_0) | \pi(x_0)$. What is the exact relation between $\pi_b(x_0)$ and $\pi(x_0)$? Are they generally equal?

10. ALGORITHMS FOR DETERMINING LENGTH OF PERIOD AND RANDOM ACCESSING

The following two theorems provide algorithms for determining (1) the period π of x_0 (the $x^2 \bmod N$ sequence that begins with x_0), and

---

[10] Roughly three fourths of all special numbers of the prescribed form satisfy this additional condition (that 2 is a quadratic residue with respect to at most one of P_1 and Q_1). The condition is needed: for example the special number in prescribed form, $N = 719*47$, fails this condition (for this N, $ord_{\lambda(N)/2}(2) = \lambda(\lambda(N))/2$).

(2) the ith element x_i.
These will be useful in the cryptographic applications.

THEOREM 8
There exists an efficient algorithm A which, when given any N of the prescribed form,[11] $\lambda(N)$, $\lambda(\lambda(N))$ AND the factorization of $\lambda(\lambda(N))$, efficiently determines the period π of any quadratic residue x_0 in Z_N^*, i.e., A[N, $\lambda(N)$, $\lambda(\lambda(N))$, factorization of $\lambda(\lambda(N))$, x_0] = π, where $\pi = \pi(x_0)$.

PROOF: To appear in the final version of this paper.

THEOREM 9
There exists an efficient deterministic algorithm A such that given N, $\lambda(N)$, any quadratic residue x_0 in Z_N^*, and any positive integer i, A efficiently computes x_i, i.e.,
A[N, $\lambda(N)$, x_0, i] = x_i.

PROOF: To appear in the final version of this paper.

Conversely, the following theorem asserts that an algorithm that knows the period, π, and the ith element x_i of the sequence $x_0, x_1,...$ obtained by squaring mod N can factor N.

THEOREM 10
Let O denote an oracle such that $O(N, x_0, i) = <\pi, x_i>$, where $\pi = \pi(x_0)$. There is an efficient probabilistic algorithm A^O such that $A^O(N) = P$ or Q, for $N = P*Q$.

PROOF: To appear in the final version of this paper.

OPEN QUESTION: Can an algorithm use an oracle that outputs just x_i -- namely, $O(N, x_0, i) = x_i$ -- to factor N?

OPEN QUESTION: Can an algorithm use an oracle that outputs just π -- namely, $O(N, x_0) = \pi$ -- to factor N?

11. APPLICATIONS

(1.1) The 1/P-generator is useful for constructing (generalized) de Bruijn sequences. These have applications, for example, in the design of radar for environments with extreme backround noise [Golomb]. We believe there may be additional interesting applications making use of properties identified in this paper, particularly the property that from 2|P|+1 but *not* |P|-1 quotient digits, one can infer the sequence backwards and forwards. For example, one could split a key, P, between two parties -- by giving |P| successive quotient digits to each so that together they have 2|P| successive digits. Neither party alone would have the slightest information *which* prime, P, was key, but cooperatively they could determine P efficiently.

(1.2) Maximum-length shift-register sequences (which are closely related to the 1/P-generator) are used for encryption of messages [Golomb]. We view the inference procedure given here as yet another step toward breaking such crypto-systems.

---

[11] $N = P*Q$, where P, Q are primes congruent to 3 mod 4.

On the other hand, we would be interested to hear of any applications which exploit the property that (1) from a string of quotient digits it is difficult to determine that string's location in the $1/P$-sequence, whereas (2) given a sufficiently long such string, one can nevertheless extend it backwards and forwards.

(2.1) The $x^2 \bmod N$ sequence can be used for *public-key cryptography:* Alice can enable Bob to send messages to her (over public channels) that only she can read. Alice constructs and publicizes a number N_A, her *public key*, with the prescribed properties: $N_A = P_A * Q_A$ where P_A and Q_A are distinct primes both congruent to 3 mod 4. She keeps private the primes P_A and Q_A, her *private key*.

Bob encrypts: Suppose Bob wants to send an n-bit message $\vec{m} = (m_1, \ldots, m_n)$, where $n = poly(|N_A|)$, to Alice. Using Alice's public key, Bob constructs a *one-time pad:* he selects an integer x_0 from $Z_{N_A}^*$ at random, squares it mod N_A to get a quadratic residue x_1, and uses the $x^2 \bmod N$ generator with input (N_A, x_1) to generate the one-time pad $\vec{b} = (b_1, \ldots, b_n)$. Bob then sends BOTH the encrypted message, $\vec{m} \oplus \vec{b} = (m_1 \oplus b_1, \ldots, m_n \oplus b_n)$, AND x_{n+1} to Alice over public channels, where $\oplus$ is the exclusive-or.

Alice decrypts: From her knowledge of P_A and Q_A, her private key, Alice has enough information to efficiently compute $x_n, x_{n-1}, \ldots, x_1$ from x_{n+1} by backwards jump (Theorem 3). From that, she reconstructs the one-time pad $\vec{b}$ and, by $\oplus$-oring $\vec{b}$ with the encrypted message, decrypts the message, $\vec{m}$.

Anyone who can reconstruct (i.e., guess with some advantage) even one bit of $\vec{m}$ from knowledge of n and x_{n+1} can thereby obtain (guess with some advantage) a bit of the one-time pad $\vec{b}$. This is impossible (by the quadratic residuosity assumption and the following theorem) if $\vec{m}$ is a randomly selected message.

THEOREM (stronger version of claim 3): Suppose N is of the prescribed type. Then the factors of N are necessary -- assuming they are necessary for deciding quadratic residuosity of x in $Z_N^*(+1)$ -- to have even an ε-advantage[12] in guessing in *poly*-time any pair (k, b_k) (i.e., any bit b_k and its location k in the sequence $b_1, \ldots, b_n$), $1 \le k \le n = poly(|N|)$, given N and x_{n+1}.

PROOF: To appear in the final version of this paper.

(2.2) Having constructed a number $N_A = P_A \cdot Q_A$ with the prescribed properties, Alice can compute $\lambda(N)$ and use it, by Theorem 9, to quickly compute $x_i = x_0^{2^i \bmod \lambda(N)} \bmod N$ (for any $x_0 \in QR_N$). This means she can use word i as address to retrieve word x_i or bit b_i efficiently -- as if the $x^2 \bmod N$ generator were a random access memory that is storing a pseudo-random sequence. [Brassard] has suggested applications, e.g., to the construction of unforgeable subway tokens, where this jumping ahead is desirable.

(2.3) As pointed out by Yao, one can convert fast Monte Carlo algorithms into almost-fast deterministic ones by replacing the use of random sequences in

---

[12] Definition: A *poly*-time procedure $\mathbf{P}[N, x_{n+1}]$ has an ε-*advantage for N in guessing a pair* (k, b_k), $1 \le k \le n = poly(|N|)$ (given arbitrary x_{n+1} selected uniformly from QR_N) if and only if $\text{Prob}[\ \mathbf{P}[N, x_{n+1}] = (k, b_k)$ for some k, $1 \le k \le n$, $|\ x_{n+1}$ is selected uniformly from $QR_N] \ge 1/2 + \varepsilon$.

such algorithms by sequences produced by a cryptographically secure generator (such as the $x^2 \bmod N$-generator): If a so-converted Monte Carlo algoirithm were to behave differently (utilizing pseudo-random sequences instead of truly random ones), this algorithm itself would become a test for distinguishing between the two types of sequences (see [Yao] for the many subtle details needed for this argument).

(2.4) Cryptographically secure pseudo-random sequence generators (such as the $x^2 \bmod N$-generator) may be viewed as *amplifiers* of randomness (short random strings are amplified to make long pseudo-random strings).

(2.5) One often uses pseudo-random sequences (rather than random sequences) because they are reproducible [Von Neumann]. For the pseudo-random sequences produced by the $x^2 \bmod N$-generator, one has only to store a short seed in order to reproduce a long sequence; one does not have to store the entire random sequence.

12. ACKNOWLEDGEMENTS

We are grateful to Andy Yao for his ideas and for his encouragement of this work. We thank Silvio Micali for pointing us to the literature on de Bruijn sequences, and for his generally helpful and encouraging suggestions. The broad relevance of the quadratic residuosity assumption to Protocol Design was first pointed out by Shafi Goldwasser and Silvio Micali. We are grateful to a number of people for valuable discussions on this work, including S. Even, A. Lempel, L. Levin, J. Plumstead, and M. O. Rabin.

REFERENCES

[1] L. Adleman, "On Distinguishing Prime Numbers from Composite Numbers," Proc. 21st IEEE Symp. on Found. of Comp. Science (1980), 387-408.

[2] M. Blum, "Coin Flipping by Telephone," in Proc. of IEEE Spring COMPCON (1982), 133-137.

[3] M. Blum and S. Micali, "How to Generate Cryptographically Strong Sequences of Pseudo Random Bits," submitted to FOCS 1982.

[4] G. Brassard, "On computationally Secure Authentication Tags Requiring Short Secret Shared Keys," in Conf. Proc. Crypto 82, 1982.

[5] L. Dickson, "History of the Theory of Numbers," Chelsea Pub. Co., 1919 (republished 1971).

[6] S. Even, "Graph Algorithms," Computer Science Press, 1979.

[7] C. G. Gauss, "Disquisitiones Arithmeticae," 1801; reprinted in English transl. by Yale Univ. Press, 1966.

[8] S. Goldwasser and S. Micali, "Probabilistic Encryption and How to Play Mental Poker Keeping Secret all Partial Information," 14th STOC (1982), 365-377.

[9] S. Golomb, "Shift Register Sequences," Aegean Park Press (1982).

[10] D. Knuth, "The Art of Computer Programming: Seminumerical Algorithms," Vol. 2, Addison-Wesley Pub. Co., 1981.

[11] W. LeVeque, "Fundamentals of Number Theory," Addison-Wesley Pub. Co., 1977.

[12] G. Miller, "Riemann's Hypothesis and Tests for Primality," Ph.D. Thesis, U.C. Berkeley (1975).

[13] J. Plumstead, "Inferring a Sequence Generated by a Linear Congruence," submitted to FOCS 1982.

[14] S. Pohlig and M. Hellman, "An Improved Algorithm for Computing Logarithms over $GF(p)$ and Its Cryptographic Significance," IEEE Trans. on Info. Theory, Vol. It-24, No. 1, (1978), 106-110.

[15] M. O. Rabin, "Probabilistic Algorithm for Tesitng Primality," J. No. Theory, Vol 12 (1980), 128-138.

[16] M. O. Rabin, "Digital Signatures and Public-key Functions as Intractable as Factorization," MIT/LCS/TR-212 Tech. memo, MIT, 1979.

[17] A. Shamir, "On the Generation of Cryptographically Strong Pseudo-Random Sequences," ICALP, 1981.

[18] D. Shanks, "Solved and Unsolved Problems in Number Theory," Chelsea Pub. Co., 1976.

[19] J. von Neumann, "Various Techniques Used in Connection With Random Digits," Collected Works, vol. 5, Macmillan (1963), 768-770.

[20] A. Yao, "Theory and Applications of Trapdoor Functions," submitted to FOCS 1982.

ON COMPUTATIONALLY SECURE AUTHENTICATION TAGS

REQUIRING SHORT SECRET SHARED KEYS*

       Gilles Brassard

       Université de Montréal
       Département d'informatique et de
        recherche opérationnelle
       C.P. 6128, Succursale "A"
       Montréal, Québec
       H3C 3J7, Canada

INTRODUCTION

    As an application of strongly universal-2 classes of hash functions, Wegman and Carter have proposed a provably secure authentication tag system.[1] Their technique allows the receiver to be certain that a message is genuine. An enemy, even one with infinite computing power, cannot forge or modify a message without detection. Moreover, there are no messages that just happen to be easy to forge. Unfortunately, their scheme requires that the sender and the receiver share a rather long secret key if they wish to use the system more than once. Indeed, the length of the key is essentially $n \log(1/p)$, where n is the number of messages they wish to be able to authenticate before having to agree on a new secret key, and p is the probability of undetected forgery they are willing to tolerate. Since they also proved that $n \log(1/p)$ is a lower bound on the number of bits required by any tag system that assures security against infinite computing power, it is clearly necessary to resort to computational complexity if we wish to have a scheme usable in practice allowing a potentially very large number of messages to be authenticated.

---

* Supported in part by Canada's NSERC grant number A4107.

The obvious complexity theory based authentication tag scheme is to use directly some form of digital signature à la Diffie and Hellman.[2] In their paper, Wegman and Carter rule this out on the sole basis that it is "easy" to break given infinite computing power. We would like to point out here several other serious weaknesses of digital signatures in this context.

i) Most proposed digital signature schemes have their claims of security based on somewhat unconvincing evidence. For instance, the security of Rivest, Shamir and Adleman's scheme[3] depends on the difficulty of factoring, since it would immediately be broken should one come up with an efficient factoring algorithm. This scheme is nonetheless perhaps not equivalent to factoring: it could be that forging signatures does not actually require factoring. More interestingly, even if the factoring problem were efficiently reduced to the problem of forging signatures (in the conventional, deterministic sense), it could still be that most signatures are easy to forge, yet factoring remains infeasible. This potential weakness was brilliantly illustrated by Shamir and Adleman's new algorithms[4-5] for breaking Merkle and Hellman's public-key cryptosystem[6]: it gives no clues on how to solve the general knapsack problem, which served as basis for the design of this cryptosystem.

ii) Other public-key cryptosystems have had their security proven equivalent (in a strong probabilistic sense) to the difficulty of solving outstanding old problems of number theory, such as factorization. In these cases, however, they either fall immediately to chosen-ciphertext attacks, which are real threats in the authentication tag context, or they offer only encryption but no signature capabilities. The first problem is exemplified by Rabin's scheme, if implemented without a hashing or one-way compression function.[7] Although the problem of chosen-ciphertext attacks vanishes if compression is used, so does the proof of equivalence with factorization. The second problem is exemplified by Goldwasser and Micali's scheme using probabilistic encryption.[8] (Notice however that Goldwasser, Micali and Yao have recently developped a signature scheme whose security is equivalent to the difficulty of factoring. This scheme is also free from the additional weaknesses listed below. Its description can be found in these CRYPTO 82 proceedings.[9])

iii) In a scheme such as Rabin's, the enemy's capability of forging signatures for even a small percentage of the messages would allow him to factor large numbers. There are no theoretical reasons, however, to believe that

the enemy could not, given a message and its signature, find
the (or a) signature for a slightly modified message.

iv) Finally, there might always be messages for which
it just happens to be easy to forge a signature.

Another possible solution to the computationally secure
authentication tag problem is through the use of a conventional cryptosystem, such as the Data Encryption Standard.[10]
For instance, a DES encryption of the message could be
hashed down to a reasonably short authentication tag.
Although this might be the most economical and efficient
solution in practice, the author feels that it is unsatisfactory because its presumed security is not founded on any
theoretical basis.

The scheme proposed in this paper gets rid of all these
objections in the sense that it is provably as strong as the
original Wegman and Carter's scheme (under suitable assumptions on the difficulty of solving some number theoretic
problems) as long as the forger is not willing to use a
ridiculous amount of time. The secret shared key, however,
can be kept reasonably short even when a large number of
messages has to be authenticated.

COMPUTATIONALLY SECURE TAGS WITH SHORT KEYS

Under the assumption that the discrete logarithm problem (index finding) is intractable, Blum and Micali have
shown how to generate cryptographically strong sequences of
pseudo random bits.[11] Yao has proved that these sequences
are so strong that they cannot be distinguished from true
random sequences by any polynomial time algorithm,[12] including Coveyou and Mac-Pherson's famous spectral test.[13]
Intuitively, this means that an enemy having exact knowledge
of a large number of the bits generated from a relatively
short seed cannot do better than toss a fair coin in order
to guess any of the missing bits.

This suggests a scheme very similar to Wegman and
Carter's, except that the secret key does not have to grow
out of proportion if a large number of messages has to be
authenticated. Let p be the probability we are willing to
accept of not detecting a forged tag, let k be an integer
not smaller than $\log(1/p)$, let M be the space of messages,
and let B bet the set of bit strings of length k. Let F
be a strongly universal-2 class of functions from M to B.[1]
For integers $i \leq j$, let $s[i..j]$ denote the bits of
inclusive rank i to j generated from seed s.

The secret key shared by the sender and receiver consists of two parts: a function f chosen among the class F, and a seed s. If m is the n-th message exchanged between them, the authentication tag is simply $a(m,n) = f(m) \oplus s[(n-1)k + 1 .. nk]$, where $\oplus$ denotes the bitwise exclusive or. In other words, both sender and receiver use their secret knowledge of s to generate the next k bits of their "pseudo one-time-pad", which they use to hide the value of the secret hashing function on the message. For some applications, it may be more convenient to transmit explicitly the current value of the message sequence number n; this would not affect security as long as the same sequence number is never used twice.

Given any sequence $\{<i,m[i],a(m[i],i)> \mid i = 1,2,...,n\}$ of properly authenticated messages, an enemy cannot produce even one triple $<j,x,y>$ such that $x \neq m[j]$ and y has a probability non negligibly better than p of being $a(x,j)$. Notice that using $i = n + 1$ in the above statement implies that an enemy could not listen until he decides to go ahead and counterfeit the authentication tag of a message of his own. On the other hand, using $i = n$, we get that an enemy could not either intercept a message that has already been correctly authenticated by the sender in an attempt to modify it somehow, compute the new tag, and pass the forged information over to the unsuspecting receiver.

As a disadvantage over digital signatures, this type of scheme does not automatically provide two-way authentication. If the existence of a trusted notary public is assumed, however, Wegman and Carter point out that it is not hard to get rid of this deficiency.[1]

APPLICATIONS FOR WHICH "RANDOM ACCESS" IS DESIRABLE

In the above proposed scheme, it was important for reasons of efficiency that the messages be authenticated sequentially, because the pseudo random bits are produced in sequential order (thanks to Yao[12]). Should the sender decide for some reason to skip messages and jump from sequence number 10 to sequence number 111, both the sender and receiver would have to compute 100k bits that are never going to be used. Things are even worse if the sender decides to use sequence numbers in an arbitrary order, which once again does not reduce security as long as the same sequence number is never used twice. (Notice that this does not follow directly from Blum and Micali, but that it does follow from Yao).

An application in which this type of sender behaviour would be desirable is if there were only one receiver (now refered to as the mint), but many independent senders (the users). In this modified context, the users would actually not know the secret key: instead they are given by the mint some magnetic card containing a unique sequence number together with an authenticated message that identifies some of their physical characteristics, such as their fingerprints. Any such user can later convince the mint that the card he owns was indeed issued for him by the mint itself. This application is similar to the Idaho Falls Zero Power Plutonium Reactor's identification procedure, except that here the validation procedure has to remain a secret.[14] More interestingly, this scheme is particularly well-suited for implementing unforgeable subway tokens if verify-only memory (VOM) is available: each token would have a distinct sequence number and an empty message (the hash function would not be used here).[15]

Clearly, for the applications proposed above, the sequence numbers would come to the receiver in an arbitrary order. This would force a validation procedure based on a sequential pseudo random bit generator to behave unefficiently. Fortunately, Blum, Blum and Shub have discovered another generator, which is capable of producing the subsequence $s[n+1 .. n+k]$ in a time polynomial in the length of s, the value of k, and the logarithm of n, instead of being polynomial in the value of n.[16] This generator can therefore be used to implement efficiently validated and provably impossible to counterfeit (under suitable number theoretic assumptions) VOM type of subway tokens.

Another potential application for this new generator is that it may enable a substantial simplification in the authentication scheme proposed in the previous section for the one-sender, one-receiver case. The use of universal-2 hashing and sequence numbers may be superfluous. Let the sender and the receiver share some secret key s for the new generator. The tag associated with message m is simply $s[(m-1)k + 1 .. mk]$, just as if the message m were taken to be its own sequence number, while the "hashing" function sends everything to zero. Unfortunately, the security of this idea does not follow directly from the work of Blum, Blum, Micali, Shub and Yao, because it uses a sparse subsequence of a sequence of pseudo random bits that is much longer than the seed. Silvio Micali and the author are currently investigating the security of this type of scheme.

HOW TO IMPLEMENT A CLUB

A club is a select group of mutually trusting members. In order to provide security against the untrusted outside world, clubs need authentication schemes with the following properties:
- any member of the club can authenticate messages,
- whenever a member authenticates a message for the benefit of some other member, he does not have to notify anyone else in the club, and
- whenever a member receives an authenticated message, he can make sure the message originates from within the club.

Moreover, the following property would be desirable:
- the amount of information that must be kept by each member is small and independent of the size of the club.

Clubs are obvious to implement using the sequence-number-free scheme suggested at the end of the previous section, assuming the latter is indeed secure: each member of the club is given a common secret key that is used for authenticating any message sent within the club. If this scheme is not trusted, so that small sequence numbers have to be used, there is a slight problem because it is very important that the same sequence number never be used more than once. One solution it to give each member a distinct small prime, in addition to the common secret key. Thereafter, whenever a member wishes to authenticate a message, he chooses as sequence number the product of the square of his private prime with some small prime never used by him before. This obviously assures unicity of sequence numbers among all numbers used within the club. The imaginative reader will easily come up with similar schemes that do not waste as many sequence numbers.

ACKNOWLEDGEMENTS

The author is very grateful to Silvio Micali, Manuel Blum and Shafi Goldwasser for insightful discussions on these topics. Similar ideas have been independently discovered by Goldwasser, Micali and Tong.[17]

REFERENCES

1. M.N. Wegman and J.L. Carter, New Hash Functions and Their Use in Authentication and set Equality, JCSS, 22:265 (1981).
2. W. Diffie and M.E. Hellman, New Directions in Cryptography, IEEE Trans. Info. Th., IT-22:644 (1976).
3. R.L. Rivest, A. Shamir and L. Adleman, On Digital Signature and Public-Key Cryptosystems, CACM, 21:120 (1978).
4. A. Shamir, A Polynomial-Time Algorithm for Breaking the Basic Merkle-Hellman Cryptosystem, in: "Proc. of 23rd FOCS Symposium, Chicago," IEEE, New York (1982).
5. L. Adleman, On Breaking the Iterated Knapsack Public-Key Cryptosystem, presented at: "AMS Workshop on Probabilistic Computational Complexity, Durham," A. Meyer, chair. (1982).
6. R. Merkle and M.E. Hellman, Hiding Information and Receipts in Trap-Door Knapsacks, IEEE Trans. Info. th., IT-24:525 (1978).
7. M.O. Rabin, Digitalized Signatures and Public-Key Functions as Intractable as Factorization, MIT/LCS/TR-212 (1979).
8. S. Goldwasser and S. Micali, Probabilistic Encryption & How to Play Mental Poker Keeping Secret all Partial Information, in: "Proc. of 14th STOC Symposium, San Francisco," ACM, New York (1982).
9. S. Goldwasser, S. Micali and A. Yao, On Authentication, Digital Signatures and Contracts in Presence of Meddler, in: "Advances in Cryptography: Proceedings of CRYPTO 82," R. Rivest, ed., Plenum Press, New York (1983).
10. National Bureau of Standards, Federal Information Processing Standards Publication no. 46.
11. M. Blum and S. Micali, How to Generate Cryptographically Strong Sequences of Pseudo Random Bits, in: "Proc. of 23rd FOCS Symposium, Chicago," IEEE, New York (1982).
12. A. Yao, Theory and Application of Trapdoor Functions, in: "Proc. of 23rd FOCS Symposium, Chicago," IEEE, New York (1982).
13. R.R. Coveyou and R.D. MacPherson, Fourier Analysis of Uniform Random Number Generators, JACM, 14:100 (1967).
14. G.B. Kolata, New Codes Coming into Use, Science Magazine, 208:694 (1980).

15. C.H. Bennett, G. Brassard, S. Breidbart and S. Weisner, Quantum Cryptography, or Unforgeable Subway Tokens, in: "Advances in Cryptography: Proceedings of CRYPTO 82," R. Rivest, ed., Plenum Press, New York (1983).
16. L. Blum, M. Blum and M. Shub, A Simple Secure Pseudo-Random Number Generator, in: "Advances in Cryptography: Proceedings of CRYPTO 82," R. Rivest, ed., Plenum Press, New York (1983).
17. S. Goldwasser, S. Micali and P. Tong, How to Establish a Private Code on a Public Network, in: "Proc. of 23rd FOCS Symposium, Chicago," IEEE, New York (1982).

Session II: Modes of Operation

SOME REGULAR PROPERTIES OF THE 'DATA ENCRYPTION STANDARD' ALGORITHM[1]

Donald W Davies

National Physical Laboratory
Teddington, Middlesex, UK

A cipher function $y = E(k,x)$ should appear to be a random function of both the key k and the plaintext x. Any regular behaviour is of interest to the users. In the extreme case regular properties might point to a weakness of the cipher. Precautions are needed in the use of a cipher that has regular features. This note describes five regular properties of the 'Data Encryption Standard' or DES, two of which have been described elsewhere, are included for completeness.

The Complementation Property

If x, k and y are related by the DES, that is $y = E(k,x)$ and if x', k' and y' are the ones complements of x, k and y respectively, it follows that $y' = E(k', x')$.

It has been suggested that if the value of key k is being sought and we have available both $y_1 = E(k, x)$ and $y_2 = E(k,x')$ then, for each key value t that we are testing, if we calculate $E(t,x)$ this can be checked for equality with both y_1 and y_2'. The former equality gives the possibility that $k = t$ and the latter gives $k = t'$. So the number of calculations is halved. In practice it is unlikely that both y_1 and y_2 will be known unless the 'chosen plaintext' criterion is applied, so this is not a serious limitation.

Nevertheless it is legitimate to ask how the property could have been avoided. One method is to remove a few of the outputs of the E 'expansion permutation' so that less than 48 digits enter the $+_2$ operation and therefore a few of the key bits enter directly into the S box inputs.

---
[1] This paper was presented on August 24, 1981, at the CRYPTO 81 conference at the University of California, Santa Barbara.

The Weak Keys

If both of the C and D registers contain either all ones or all zeros, their outputs will be constant and the sequences of keys Kn will be repeat of the same value 16 times. Consequently, reversing the sequence of Kn has no effect, which implies that encryption and decryption are the same operation. For such a key k,

$$y = E(k,x) \text{ implies } x = E(k,y), \text{ ie } y = D(k,x).$$

These are called 'weak keys' and there are four of them, shown in the table in hexadecimal notation for the four choices of C and D register contents. There are two complementary pairs:

C	D	Weak Key Values			
Zeros	Zeros	0101	0101	0101	0101
Ones	Ones	FEFE	FEFE	FEFE	FEFE
Zeros	Ones	1F1F	1F1F	0E0E	0E0E
Ones	Zeros	E0E0	E0E0	F1F1	F1F1

This feature could have been avoided by making the C and D registers a little longer and placing suitably chosen bit values in the extra places - in effect a fixed extension of the key. This would also have avoided the complementation property.

The Semi-weak Keys

The shift pattern for encipherment is:

    1 1 222222 1 222222 1

The first key K_1 is taken after the first shift. The last key K_{16} is taken after the last shift, when the contents of registers C and D are back at their starting positions again.

For decipherment it follows that the shift pattern is (right shift):

    0 1 222222 1 222222 1

where the first key, K_{16}, is taken without shifting and the last key when the contents of the registers are one shift from home.

If the pattern 0101010101010101010101010101 is used in one of these registers for encipherment it has only 2 values under all the shifts. The same key sequence appears when 1010101010101010101010101010 is used for decipherment, because:

(a) The extra shift at the start in encipherment requires the two patterns to be different.

(b) Apart from this initial shift, the shift pattern is palindromic.

No other pattern will produce this result. Its practical effect can be stated as follows. There are two keys s and t such that:

$$y = E(s,x) \text{ implies } x = E(t,y), \text{ ie } y = D(t,x).$$

To arrive at a list of all these pairs of keys, let the 0101 .... pattern be called U and the 1010 .... pattern be called V. Six pairs of semi-weak keys exist.

In the table of semi-weak keys, the first two pairs contain complementary keys. Among the other 4 pairs, one pair is the complement of another pair.

C	D	Semi weak Key Values			
U	U	01FE	01FE	01FE	01FE
V	V	FE01	FE01	FE01	FE01
U	V	1FE0	1FE0	0EF1	0EF1
V	U	E01F	E01F	F10E	F10E
U	Zeros	01E0	01E0	01F1	01F1
V	Zeros	E001	E001	F101	F101
U	Ones	1FFE	1FFE	0EFE	0EFE
V	Ones	FE1F	FE1F	FE0E	FE0E
Zeros	U	011F	011F	010E	010E
Zeros	V	1F01	1F01	0E01	0E01
Ones	U	E0FE	E0FE	F1FE	F1FE
Ones	V	FEE0	FEE0	FEF1	FEF1

This feature of semi-weak keys could have been avoided by choosing an asymmetric pattern of shifts, for example:

```
Encipher    1 1 22222 1 2222222 1 (left)
Decipher    1 2222222 1 22222 1   (right)
```

But even with this precaution, the keys would be weak in the sense that the keys K_1 ...... K_{16} take only two different values. In the same way it could be argued that keys resulting in patterns with period 4, such as 00010001 ... in the C or D registers are relatively weak because only 4 different keys are produced. Perhaps a better method of avoiding these weaknesses is to extend the key with some fixed input, such as three bits 001 in each of C and D, which become 31 bit registers.

The effect of 'key weakness' on the 16 rounds of encipherment

If a pattern of successive K_n values such as abcdxdcba occurs, the reversed part of the sequence, dcba, produces a transformation of the L and R registers which is the inverse of the abcd operation. The intervening x operation is essential (in the inversion of DES itself, x is represented by the extra interchange of L and R at the end of the rounds). The result of the sequence described is that the x operation is subjected to a 'projection' of the kind $P \times P^{-1}$.

Viewed in this way, the weak keys which produce a Kn pattern aaaaaaaaaaaaaaaa take the form $aPaP^{-1}$ or $PaP^{-1}a$. The semi-weak keys give a Kn pattern abbbbbbbaaaaaaab which is of the form $aPbP^{-1}QaQ^{-1}b$. The keys with period 4 in the C and D registers (of which there are 240 in addition to weak and semi-weak) produce a Kn pattern abdbdbdbcacacacd which is of the form $aPdP^{-1}QaQ^{-1}d$, very similar to the form of semi-weak keys.

The effect of this on the cipher is unknown, but it suggests that in sensitive applications the 240 keys of period 4 in the C and D registers should be avoided. No other such patterns are known to the author. It is true that other periods can occur in the C and D register, such as period 3. There are only 3 different values of Kn in this case, but their use is grouped so that it does not introduce a new structure into the operations, as far as we can tell.

The Permutations P and E in relation to the S-Boxes

The strength of the DES lies in the complexity of the function resulting from iteration round the loop (R)E S P (R), where (R) represents the contents of the R register and E the expansion permutation, S the S-boxes and P the permutation of that name.

The appearance of the DES can be changed without materially affecting it, for example the IP permutation is of this kind. Since an arbitrary re-numbering of the bits of both L and R is unimportant, it is not the individual permutations P and E that are significant but their product, the operation P followed by E.

The identification of the S-box inputs and outputs is also mostly insignificant. All the outputs are of equivalent status. The inputs, due to the nature of E, divide naturally into 3 classes. Let us denote the six input bits of an S-box as abcdef. Then the ef values of one S-box equal the ab of its neighbour to the right, for example in the case of S-boxes 1 and 2, $e1 = a2$ and $f1 = b2$. For our purpose, input bits a and b are not distinguishable because a permutation of the bits of the L and R registers could interchange any pair of these, interchanging also the corresponding ef pair. The c and d inputs of any S-box is also indistinguishable.

The significant feature of the PE permutation is therefore not the bits they interconnect but the groups ab, cd and ef which they join. The table shows for each of the S-box outputs which S-box input it affects, by way of the PE permutation. For example, S-box output bit 1 of S1 affects f2 and b3 on boxes S2 and S3 respectively.

Some Regular Properties of the DES Algorithm

S1	S2	S3	S4
f2 f4 d6 d8	f3 e7 c1 c5	e6 e4 c8 c2	c7 e5 c3 f8
b3 b5	b4 a8	a7 a5	a6 b1

S5	S6	S7	S8
e2 c4 f6 d1	e1 f7 d3 d5	e8 e3 c6 d2	f1 d7 d4 f5
a3 b7	a2 b8	a1 a3	b2 b6

The way in which the outputs have been arranged in the DES obscured the pattern which is already evident, namely that of each 4 output bits, two go to c,d inputs and two go to a,e or b,f inputs. The pattern can be shown best as a graph with directed arcs pointing from S-box output to S-box input. We show graphs separately for the c,d and e,f inputs in Figure 1. The nodes of the graphs are the eight S-boxes.

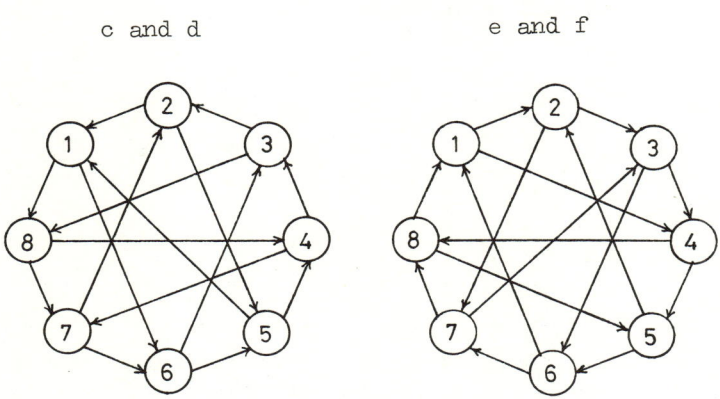

Figure 1. Graphs showing the influence on the S-box inputs from their outputs via the E and P permutations.

The graph cd can be expressed as two Hamiltonian cycles (87654321) and (84725163). The graph ef has an identical form reflected about the vertical line. Both are very regular graphs which seem likely to have been a design intention. The apparent randomness of the P permutation is due only to the way that S-box inputs and outputs have been identified.

The corresponding graph for the ab inputs shown in Figure 2 is different in form and much less regular in appearance. It is a consequence of the ef graph because of the a,e and b,f equalities due to the expansion in the E permutation. S4 has a special position and it is the most structured of all the S-boxes.

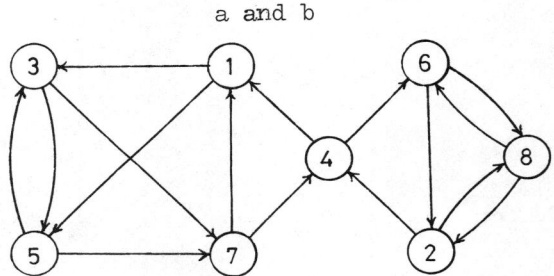

Figure 2. Continuation of Figure 1 showing the influence on a and b inputs.

The cd graph can be drawn with the other Hamiltonian cycle on the outside, which presents a new picture shown in Figure 3 (and likewise the ef graph).

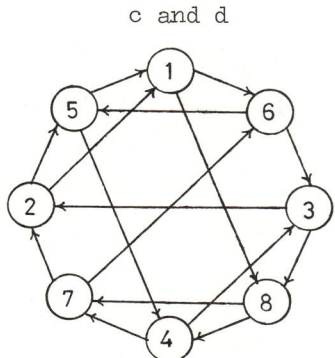

Figure 3. Repeat of part of Figure 1 in a different arrangement.

We can now see that alternate S-boxes in the outer cycle belong to 1234 and 5678. Hellman noted a difference between these sets in the structure of the rows of the S-boxes.

Interpretation of the Regularity of Permutations

This regularity is one of construction. It does not, to our knowledge, result in any regularity of the DES function. The purpose of the permutations is to spread any change among the S-boxes in as few rounds as possible. For this purpose the ef and cd graphs are well chosen. For example, a change in the input of S1 spreads to 6 and 8; from then on the S-boxes affected in successive rounds are 3457, then 1234678, by which time all S-box inputs have been affected — in three rounds only. Since only half of the 64 bits circulating in the LR registers are changed at each round, 6 rounds are needed to spread a single bit change completely. The chosen graphs cd and ef are very efficient in this respect.

Any weakness due to the regularity would, if it were present, be due to a corresponding regular feature of the S-box functions. No such regularity has been found.

Conclusions

In this note we have described four functional regularities of the DES algorithm and one regularity in its structure.

The regularities of function could be avoided by a change in the key generation part of the DES algorithm. For example, each of the C and D registers could be extended from 28 to 31 bits in length. When loading them from the 56 bit key, three bits in each register would be set to fixed values, say 001.

The regularity of the structure of permutations might have been a design feature, intended to spread any single bit change as quickly as possible.

There is no evidence that any of these regularities produces a weakness of the algorithm. The weak keys can be avoided and are therefore just a nuisance. The design of the algorithm could have avoided these regular features.

THE AVERAGE CYCLE SIZE OF THE KEY STREAM IN OUTPUT FEEDBACK ENCIPHERMENT

Donald W Davies and Graeme I P Parkin

National Physical Laboratory
Teddington, Middlesex, UK

SUMMARY

Output feedback is a method of using the 'Data Encryption Standard' (DES), and it is defined in Federal Information Processing Standards Publication 81 produced by the US National Bureau of Standards. Its purpose is to provide a method of using the DES without error extension, since both encipherment and decipherment consist of adding a 'key-stream' modulo 2 to the data stream. It has been defined for the values m = 1, 2, 3 .... 64 of the parameter m, which is the feedback width.

This kind of stream cipher depends for its security on the unpredictability of the key-stream. Being a finite state machine with 2^{64} states it must enter a repetitive cycle of 2^{64} states or less. If a cycle can be found (ie there are two identical segments of key-stream of non-trivial extent) messages M_1 and M_2 enciphered with identical segments provide, from the modulo 2 addition of their ciphertexts, the value of $M_1 + M_2$, and this allows M_1 and M_2 to be discovered if there is enough redundancy. Therefore repetition of the key-stream must be prevented, which implies that the stream must not be allowed to continue long enough, with a given key, that a repeat has more than a very low probability. There are similar dangers when one key value is used for an excessive amount of data even if the generator is re-initialised from time to time by randomly chosen IVs. Therefore a good key-stream generator should have an average cycle length which is very large, preferably comparable with 2^{64}, the maximum value it can have for the OFB key stream generator.

In this paper we calculate the average cycle lengths for each value of m. The case m = 64 is a well known result in the theory of permutations with average cycle length $2^{63} + \frac{1}{2}$. For the other values, an estimate of average cycle length can be made only by using an approximate mathematical model of the generator which is rather difficult to justify. The average is $(\pi n/8V)^{\frac{1}{2}} + \frac{1}{3}$ where, for small m, $V = 1 - 2^{-m}$ and for large m, $V = 1 - 2^{m-64}$. Because of the doubtful assumptions in our model we have carried out an experiment with 8 bit register in place of R and randomly chosen permutations of the 256 states in place of DES. The results of the experiment confirm the theory and encourage us to apply it to the actual OFB configuration. The broad conclusion is reached that OFB with m = 64 is reasonably secure but OFB with any other value of m is of greatly inferior security. Since OFB with values of m other than 64 has no advantages we propose that the only recognised OFB mode of operation should be m = 64.

ANALYSIS OF CERTAIN ASPECTS OF OUTPUT FEEDBACK MODE

Robert R. Jueneman

Satellite Business Systems
8283 Greensboro Drive
McLean, VA 22102

0. ABSTRACT

The Output Feedback (OFB) mode of operation of the Data Encryption Standard (DES) is discussed, and compared to the other DES modes. The advantages of the Output Feedback mode's insensitivity to transmission errors and the applicability to bulk encryption of multiple users' transmissions are presented, along with the disadvantages of an increased sensitivity to bit slippage and a requirement for more complex synchronization procedures.

It is concluded that the Manipulation Detection Code technique suggested in draft Federal Standards 1025 and 1026 is unsound, and that therefore there are only differences of degree in the vulnerability to active (spoofing) attacks between the various modes. Two separate encryption operations are required to provide cryptographic protection against both the passive and the active threat, but a quadratic residue checksum is proposed as a possible alternative. However, considerations of the physical media involved and the types of traffic carried may make even this level of protection unnecessary for many applications.

The problem of transmission in depth is discussed, and Output Feedback mode is analyzed with respect to the probability of repeating a given output prior to exhausting the space of 2^{64} variables. Reiterating the advice of Davies and Parkin, the user is cautioned not to use K<64 bit feedback and it is recommended that FIPS PUB 81 be revised to delete that option. Numerical data are presented for various reinitialization rates which indicate that when OFB is used not more than four billion iterations or 10,000 reinitializations or one day of operation should occur between DES key

changes. One week to one month between master key changes
is suggested, especially for cryptographic networks of more than
two stations. Blakley's shadow key concept is recommended as a
way of minimizing the possibility of human compromise.

Appendices discuss the existence of 256 weak, semi-weak, and
demi-semi-weak keys, plus the derivations of the formulas for the
probability of repetition for the various cases.

KEY WORDS: Data Encryption Standard, DES, Output Feedback mode,
non-error multiplicative ciphers, data-independent ciphers, active
attack, spoofing, Manipulation Detection Codes, DES cycle length,
transmission in depth, crypto period, weak keys, semi-weak keys,
demi-semi-weak keys, cryptographic synchronization, key change
schedule.

1.0 INTRODUCTION

The publication of FIPS PUB 81 by the National Bureau of Standards[1] added another mode of operation, the so-called Output Feedback (OFB) mode, to the list of "approved" modes of operation of the Data Encryption Standard[2] block cipher. Because DES is now being used extensively to encrypt voice and other non-data applications, it is expected that OFB will soon become the most important mode of operation by far, at least in terms of the amount of traffic carried and probably the dollar volume of units sold.

It may be of some interest to know why Output Feedback mode would be used as opposed to one of the more conventional modes, for it has both advantages and disadvantages. If we compare it to the other three DES modes, i.e. Electronic Code Book (ECB), Cipher Feedback (CFB), and Cipher Block Chaining (CBC), we can see that those modes have a significant disadvantage from the standpoint of error propagation. If the communication channel is not error free, then the use of DES will cause many more bits to be received in error than would otherwise be the case.

In the case of ECB mode, a one bit error in transmission will cause on the average fifty percent of the plaintext bits of the decrypted block to be in error. The same is true of CBC, but in addition, the following block will have the same number of bits in error, and in the same position, as the transmission error bits. The CFB mode preserves bit identification, so the same bit that is received in error will be wrong in the plaintext, but succeeding bits will be in error until all of the transmission errors have been shifted out of the DES input block.

In today's typical data transmission protocols such as bi-sync, SDLC, or HDLC, an ARQ (automatic repeat request) or Go-Back-N discipline is used in combination with a cyclic redundancy check to validate or reject each message or message portion that is received. The error extension characteristics of EBC, CFB, or CBC thus do no harm, and may reduce the possibility of undetected errors.

However, for other applications such as voice, facsimile, or video, the use of one of these other modes would seriously degrade what is often a very precious commodity -- the system's signal-to-noise ratio. If the communications channel were almost perfect, then the effect would probably be tolerable. But if the channel were occasionally noisy, encryption might make effective communication impossible.

For example, if a 32kbps delta modulation technique were used to encode voice signals (by using alternating ones and zeroes to indicate a constant level, two ones to mean "louder", and two zeroes to mean "softer"), then the effect of noise is to make the short term loudness level rather erratic, as though someone were fiddling with the volume control on a radio. But although the effect of the noise may be objectionable, it does not significantly impair intelligibility until a bit error rate of around 1% is reached. Obviously, however, if decryption were to transform that 1% error rate into a 32% error rate voice communication would become impossible. The relative impact of a decrypted error upon a conventional 64kbps PCM digitized voice channel would be even worse.

For this reason, Output Feedback mode is preferred for such applications as discussed in[3,4]. By emulating the classical Vernam stream cipher wherein a keystream is exclusive ORed bit by bit with the plaintext, the impact of a single bit transmission error in the ciphertext is limited to the corresponding bit in the plaintext after decryption. Because the keystream is generated internally (as we shall see), it is presumed to be error free, and OFB is usually referred to as non-error multiplicative.

Because the OFB keystream does not depend upon any previous data in the generation of its keystream output, it is also called data-independent. If a transmission error does occur, or if an active attack were to succeed in modifying one or more bits of the output, that fact could not be detected through the use of OFB alone. Indeed, one of the potential advantages of OFB is that the processing time necessary to encrypt or decrypt the traffic can be overlapped with or even preceed the generation or reception of the traffic itself, so long as the encryption/decryption processes are properly synchronized.

As will be discussed, OFB has two potential shortcomings which mitigate against its indiscriminate use. First of all, the non-error propagating nature of OFB means that by itself it cannot provide any cryptographic protection against the so-called active or spoofing attacks, since the data independent nature of OFB would allow an attacker to add, drop, repeat, or rearrange messages without detection unless other precautions are taken. This problem will be discussed further in section 2.1, where it is concluded that the techniques commonly proposed to detect active attacks are unsound in certain cases, and that the difference in vulnerability to active attack between the various modes is therefore primarily one of degree. A quadratic residue checksum technique is proposed that may provide sufficient protection against spoofing for most applications. It is also concluded, however, that in important instances the nature of the transmission medium is such that an active attack would be exceedingly unlikely if not physically impossible, and that other defenses are available as well.

Second, although OFB is relatively insensitive to transmission bit errors, it is quite sensitive to clocking errors or bit-slippages. Whereas Cipher Feedback is self-synchronizing after 64 bits and Cipher Block Chaining is usually implemented with a message structure which indicates the start of a message in the clear (with a beginning of text - end of text indicator), Output Feedback mode requires rigid timing and synchronization from some external source. If the bit timing is ever lost, all subsequent text is garbled until a resynchronization occurs. With respect to the bit slippage problem, however, modern multiplexing systems have to guard against that problem in any case. The problem will be discussed further in section 4.0, but suffice it to say that a variety of solutions are available for this problem as well.

As a result, Output Feedback mode is particularly well suited to the task of bulk encrypting multiple voice, data, and/or video connections being carrried via satellite or microwave links. The OFB mode can be made to operate at very high speeds; it is transparent and insensitive to the traffic being carried and does not require any special headers, etc.; it does not degrade the signal-to-noise ratio of the system; and in many cases, the physical characteristics of the broadcast medium can provide quite adequate protection against active attacks.

2.0 OUTPUT FEEDBACK (OFB) MCDE

The operation of the OFB mode is straight-forward. Referring to the definition of OFB in FIPS PUB 81 (Figure 1), the sender and receiver must arrange to have synchronized Initialization Vectors (IVs), either transmitted or internally generated. The DES engine operates in encrypt mode at both sender and receiver. The keystream is exclusive ORed with the plaintext to produce the ciphertext, and at the receiver the identical, synchronized keystream is again XORed against the data (in this case, the received ciphertext) to cancel out the keystream and produce the original plaintext.

Analysis of Certain Aspects of Output Feedback Mode

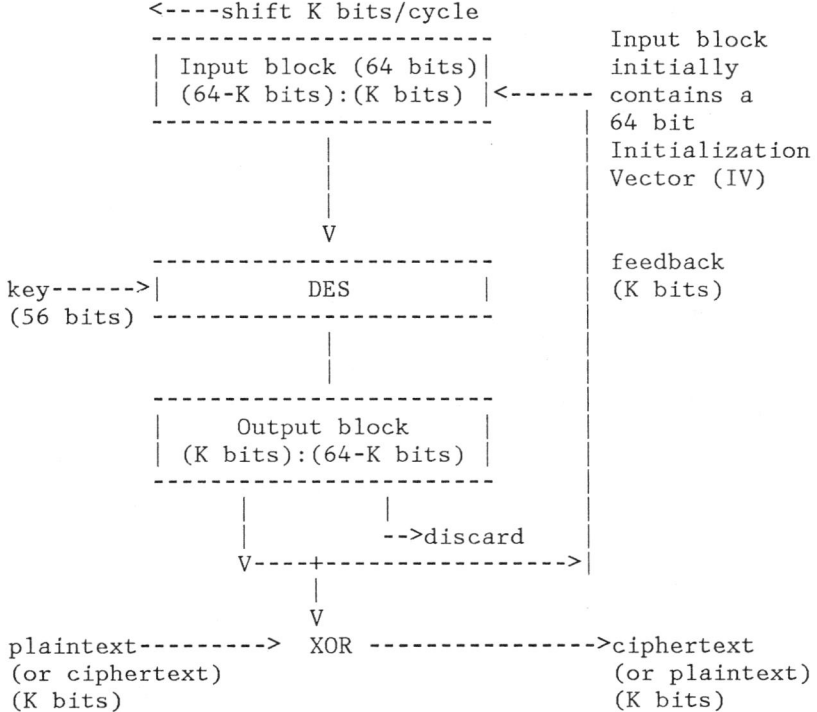

Figure 1. Output Feedback (OFB) Mode

Since the cipher keystream is generated internally, it is not subject to transmission errors. If the ciphertext is received in error, only the corresponding bits of the decrypted plaintext will be in error. For this reason, OFB is called non-error multiplicative.

Although FIPS PUB 81 describes Output Feedback mode using the above figure, it appears that the motivation for the generalization of OFB to include the K<64 bit option was for compatibility with Cipher Feedback mode, perhaps to simplify chip design. However, the analysis presented at CRYPTO 82 by Davies and Parkin[5] indicates that the use of OFB with less than full block (K=64 bits) feedback will give rise to unacceptably short cycles and the transmission in depth problem that will be discussed below. The other disadvantage of course is that the DES chip speed is wasted by throwing away bits in this configuration. Because OFB is not self-synchronizing (unlike CFB), there is little reason to use K=1, K=8, or any other value of K less than K=64. It is the author's understanding that NSA and NBS are aware of the difficulty and intend to revise FIPS PUB 81, but that process might well take a year or more. IN THE

MEANTIME, IT IS STRONGLY RECOMMENDED THAT ONLY K=64 BIT OUTPUT FEEDBACK BE USED.

2.1 Defenses Against the Active (Spoofing) Attack

Without going into unncessary detail, a necessary and sufficient set of conditions to protect against an active (message modification) attack is to provide and check for unique message segment numbers or other indicator of "time", and to protect the contents of the message along with the message segment number with a checksum or hash total which will detect any changes (including ordinary modification or "bit twiddling", as well as transposition, replication, insertion, or deletion of all or portions of the message) with a high probability. The checksum must be recomputed and verified after reception (and decryption, if necessary).

Unfortunately, not just any checksum technique will do, and the Modification Detection Code (MDC) technique contained in the current drafts of the proposed Federal Standard 1025 and 1026[6,7] proves the point. As defined, they compute a simple 16-bit continuous XOR sum, encrypting it to form the MDC. (In the most recent draft the author has seen, the MDC is extended with 8 one bits, to form a 24-bit field.) However, since CFB and CBC are self-synchronizing, portions of the message could be interchanged arbitrarily on 64-bit boundaries (or 16 bit boundaries, in the case of Cipher Feedback) without detection using that checksum technique. Since the length of the message is not explicitly indicated, the message could also be extended by replicating any 64-bit portion of the ciphertext any even multiple number of times (to ensure that the XOR product would cancel itself), then inserting the fradulent text on any two identical 64-bit fields (not necessarily adjacent) that happened to start and end on 64 bit (or 16 bit) boundaries. Thus, $10000.00 could be converted to $1.00 or to $100000000.00 with equal facility! OBVIOUSLY THIS PROBLEM WILL HAVE TO BE FIXED BEFORE THOSE STANDARDS CAN BE ADOPTED.

The most secure way of ensuring that an active attack can be detected that is known to the author is to compute a Message Authentication Code (MAC) while encrypting using Cipher Block Chaining (as defined in FIPS PUB 81) and to append the MAC to the message unit. By continuously chaining the encryption operations together, every bit of the input data is mixed with all of the other bits of the data and all of the bits of the key as well, so a single bit of difference in the plaintext would produce a veritable explosion of error bits throughout the rest of the message if all of the ciphertext were output during this process (which it is not). The MAC should then be formed by taking the high order N bits of the final DES output block and appending it to the plaintext. A MAC can also be formed by using Cipher Feedback, but it is then necessary to run the DES for one more cycle, then take the high

order bits of that output block as the MAC. An alternative
approach would be to append N bits of zeroes to the original plaintext, and to record the ciphertext corresponding to those bits as
the MAC. Regardless of which of the two techniques is used, if the
key is kept secret and the MAC is recomputed and verified after
the message is received or retrieved (after decryption, if necessary), the probability of a spurious message being accepted is
2^{-N}. (N should be at least 32 bits. If the MAC is used to protect
cryptographically sensitive operations such as key loading, N should
be a full 64 bits long.)

It might be thought that the process of forming the MAC could be
combined with the encryption used to produce the cipher text, and
thereby avoid the necessity of two encryption operations on the same
text for those situations where both privacy and protection against
message modification are desired. Unfortunately, as Miles Smid of
NBS pointed out to the author, if two extended fields of the ciphertext produced by CBC were interchanged, 64 bits of the deciphered
plaintext would be garbled at the beginning and ending of the two
fields, but the rest of the message would be unaffected. Worse yet,
the errors introduced into the plaintext in this case are precisely
the errors needed to cause the MAC to be computed in such a manner
as to be equal to the original MAC, thereby producing a false verification! This does not happen if the message and the MAC is sent
in the clear. It is a problem only if a MAC is computed on the
basis of a plaintext that resulted from the decryption of transposed
ciphertext blocks.

As a result, we have a situation similar to the use of public
key crypto algorithms. If we wish to protect against modification
only, one encryption operation is sufficient (form a MAC). And, of
course, if only privacy is desired, one encryption operation will
also suffice (form the ciphertext). For human readable information,
the garble caused by transposing ciphertext blocks may be sufficient
to detect a spoofing attack, because of the redundancy of natural
language and the syntactic and semantic rules of the language. But
if machine readable text consisting of pure binary information is to
be guaranteed protection against any possible manipulation, THEN TWO
SEPARATE ENCRYPTION OPERATIONS ARE NECESSARY. First, the MAC must be
formed, and appended to the message text to provide authentication.
Next, the message plus the MAC must be encrypted (using any desired
mode) to provide security. It is suggested that two different keys
(or one key and a variant such as 32-bit circular shift) be used for
the two different processes, although it might suffice to merely use
different IV's.

As a result, it can be stated that there is no essential difference between Output Feedback mode and the other two modes insofar
as manipulation detection is concerned. The MDC technique proposed
in the current drafts of Federal Standards 1025 and 1026 is unsound,

and although OFB might be susceptible to certain additional forms of modification, the matter is primarily one of degree. If full security and authentication is desired, two encryption operations using two different keys or at least two different IV's will be required.

However, if both privacy and protection against an active attack are considered necessary but the use of double encryption to form a MAC is undesirable, an alternative to the exclusive OR technique may provide adequate protection; albeit with some message overhead and without quite the same cryptographic strength. A quadratic residue checksum generator of the form

$$Y(i+1) = (X(i) + Y(i)^2) \text{ MOD } C, \quad i = 0 \text{ to } n$$

is suggested, where $C=2^{31}-1$, $Y(0) = 0$, $X(0)$ is a secret, random 32-bit seed that changes every message and that is transmitted as a prefix to the message, $X(1)$ to $X(n)$ are 32-bit substrings of the message (padded on the right with the high order bit of the seed $X(0)$ as necessary); and $Y(n+1)$ is the final checksum. It is assumed that the squaring, addition, and division operations are carried out in a 64-bit register without overflow. The important aspects of this checksum technique are that it is non-linear and not commutative; it involves every bit of the message in a position sensitive manner; and the use of a secret, random seed makes it impossible to predict the result, even in the case of a known plaintext attack against Output Feedback mode. The use of a unique Chain Identifier (CID) within the message text is still necessary to protect against the replay, substitution, or deletion of an entire message chain.

2.2 Physical Media Considerations

This degree of protection against OFB spoofing attacks may be unnecessary, however, for it is often the case that the physical characteristics of the transmission medium are such that although it may be possible to jam the transmission or interrupt the communications at will, it would not be possible to spoof the traffic without the interruption being quite obvious.

For example, it might well be possible to intercept information via a leakage path in the case of fiber optics, coaxial cables, or millimeter wave guides; but inserting traffic without breaking and splicing the medium in a detectable manner would be quite difficult. Similarly, although it would be feasible to receive a transmission broadcast via satellite or even jam the satellite or ground receiver, it would be virtually impossible to covertly absorb or block the signal from the transmitting earth station to the satellite. And even if that could be done, it would still be necessary to analyze the outgoing traffic in real time to know what should be replaced, and then broadcast a fraudulent replacement signal without

detection. Merely broadcasting a stronger signal would not spoof
the system, for the combination of the two signals (one legitimate
and the other one not) at the satellite would cause observable power
output differences and probably result in harmonic distortion
products or "spatter". In addition, the timing would have to be
synchronized to the bit, which would be almost impossible without
observable closed loop tracking because of the complexities of the
satellite motion and the consequent variations in the round trip
delay of the signal.

The fundamental goal of all cryptographic systems is to make
penetration of that system prohibitively expensive COMPARED TO THE
VALUE OF THE INFORMATION BEING PROTECTED. There may be many cases,
therefore, where the use of OFB to provide secrecy, plus the physical
characteristics of the transmission (or storage) medium, plus the
relatively unlikely possibility of an active attack in the first
place will provide protection that is quite sufficient for the given
application.

3.0   THE CRYPTOPERIOD OF OUTPUT FEEDBACK MODE

3.1   The Transmission in Depth Problem

It should be noted that any data independent mode of operation
could be broken instantly by a known plaintext attack if the same
series of encryption bits were ever used twice.

In fact, if the cipher stream itself repeats, or if the keystream is inadvertently reused to encipher two different plaintexts
it may be possible to break the enciphered text WITHOUT RECOVERING
THE KEY, AND WITHOUT KNOWING THE PLAINTEXT OF EITHER MESSAGE. Consider the case where two message streams have been encrypted using
the same stream of DES output bits, but perhaps at different points
in time. By systematically XOR'ing the two ciphertext streams
together at different points and then examining the resulting text
with the standard statistical tests for correlation,[10] the common
point of occurrence in the DES output stream could be discovered.
The XOR of the two ciphertexts would cancel out the DES keystream,
leaving as the result the XOR of the two plaintexts. Assuming that
the plaintexts have the built-in redundancy of any natural language
(including speech, most computer output, and any coherent video
image), this type of running key cipher can be solved by techniques
that have been known for more than sixty years[8]. This general
problem is called "transmission in depth".

3.2 Weak Keys and Potentially Weak Keys

Since data independent modes of operation are vulnerable to any transmission in depth, questions regarding the length of the sequences generated by the Output Feedback mode and the possibility of "short" sequences would seem to be quite appropriate; especially since four "weak" keys are known to produce cycles of length two. In addition, as discussed in Appendix A, there are other keys which are referred to as semi-weak, demi-semi-weak, and hemi-demi-semi-weak because of their structure. Although those keys (as many as three billion of them) are not known to cause a short OFB cycle, they have not yet been adequately investigated to the author's knowledge.

(It may be worth restating the conclusion of Appendix A, to ensure that it is not missed by the casual reader: AVOID KEYS WITH AN OBVIOUS STRUCTURE, ESPECIALLY THOSE MADE UP OF 0's, 1's, E's, OR F's PREDOMINATELY, by using random or pseudo-random generation techniques, or by screening the keys as they are generated.)

3.3 Analysis of OFB Cycle Length

The DES algorithm does not, of course, generate genuinely random output values as a function of the input. If it did, decryption would be impossible, since the mapping from the plaintext space to the ciphertext space would not be one-to-one. (However, the sequence of bits generated by the Output Feedback mode is apparently very close to a random stream of bits with regard to its statistical properties, although to the author's knowledge no detailed tests of randomness along the lines discussed by Knuth[10] have been documented in the literature with the exception of a now rather old analysis of the power spectrum[11].)

Because DES has a unique inverse, all possible points out of the space of 2^{64} variables must each be a member of one and only one disjoint cycle, at least if we assume a constant key and K=64 bit feedback. Because of this fact, the actual randomness of the individual bits output by OFB is not as important as the fact that we can model the process as a random selection of "objects" in a permutation cycle. Although there is no known way (other than a trial) to determine the number of cycles that a given key will produce; nor the minimum, maximum, or average length of a cycle produced by a given key, it is possible to calculate the probability of a repetition as a function of the number of iterations, and the average cycle length as well, as averaged over all possible keys assuming the random selection process.

Appendix B presents the analysis of the cycle length of 64-bit Output Feedback mode for the cases of pure OFB, OFB with very frequent random or psuedo-random reinitializations, and the more general case of Y random reinitializations after X iterations. For the case where the IV is never reseeded, the probability of a repetition would be approximately $N/2^{64}$ and the average cycle length approximately 2^{63}. On the other hand, if the IV were randomly reseeded after every iteration, the probability of a repetition would be approximately $N^2/2^{65}$ and the average cycle length approximately 2^{31}. The general case involving periodic IV reseeding operations is also presented, with formulas and numeric data.

(Davies and Parkin[5] have independently confirmed and extended this analysis to cover the case of non-invertible modes of DES, particularly for K<64 bit feedback, concluding that in that case the average cycle length is on the order of 2^{31} and that there is typically a dense set of "trees" or sequences that eventually converge to a very small number of distinct cycles (approximately 22), also of length 2^{31}. Also presented at CRYPTO 82 was the paper by Hellman and Reyneri[12], who discussed the case where the output was fed back into the key instead of the input, obtaining results similar to that of Davies and Parkin.)

3.4 Alternatives to Output Feedback Mode

It could be argued that it would be preferable to use no feedback at all, but simply drive the DES engine with the output of a function which has a guaranteed long period, such as an incrementing counter. The problem of reinitialization could be taken care of by using some of the high order bits as a counter for the reinitialization, with the low order bits being used in lieu of the output feedback to generate the keystream. We would therefore eliminate the very low probability of a short cycle occurring.

But the author's intuitive feeling is that to deliberately expose any cryptosystem to a known systematic input such as an increasing counter would represent an unnecessary risk. The essence of a good cryptosystem lies in increasing the entropy of the information at every possible point, and encrypting a counter does not appear to increase the entropy nearly as much as encrypting the output of DES. In addition, an incrementing counter might allow some form of a pre-computation attack, because it suffers from predictability.

Perhaps someday something like the power spectrum approach will be applied to the cryptanalysis of OFB, or other techniques might yield results that are presently unforeseen. While there may be no known means of mounting a cluster attack on DES, it does not seem wise to do part of the attacker's work for him.

Some other modifications might be possible, including extending the shift register to more than 64 bits, but the Davies and Parkin type of analysis would have to be considered before this could be recommended, and there does not appear to be any particular advantage associated with such a change. Similarly, some have suggested that the feedback path be made more complex, with complicated non-linear elements, maximal length shift register sequences, Fibonacci generators, etc. However, the author's experience cautions that true improvements do not come easily, and that there are a surprising number of ways to make mistakes that are not at all obvious. To make changes without clearly understanding all of the consequences runs the risk of weakening, rather than strengthening the algorithm. The difficulty with the approved K<64 bit feedback configuration and the MDC technique proposed in the draft Federal Standards 1025 and 1026 illustrates this principle quite well.

Finally, there is a good, practical reason for using OFB instead of an incrementing counter, at least for certain applications. The astute reader may have noticed that the keystream generated by K=64 bit Output Feedback mode would be precisely the same as that produced by Cipher Block Chaining (CBC) mode if the data input to CBC were a constant string of zeroes. As it happens, the AMD crypto chip implements CBC mode on the chip, so a very fast and simple way to obtain OFB mode is to insert zeroes as "data" and run in CBC mode, externally exclusive ORing the resulting keystream with the plaintext/ciphertext as the keystream is being extracted. An incrementing counter would require the use of ECB mode, which would be somewhat slower.

4.0 OUTPUT FEEDBACK SYNCHRONIZATION

A potentially significant problem with the OFB encryption mode is that it runs open loop. If somehow the transmitter and receiver fall out of synchronization, the system may not automatically recover unless the end user can detect that a synchronization failure has occurred and somehow cause a system reset. With the layered architecture data transmission protocols that are increasingly coming into use, this can lead to awkward implementations. In the case of uni-directional broadcasts, by definition no reverse flow messages occur and all of the text would be lost following a bit slip or loss of sync.

In real-life systems that are subject to jitter, clock slippages, occasional outages, etc., this can be a serious problem. One possible solution is, of course, to resynchronize the system at predetermined intervals, in order to minimize the amount of outage time. For example, if OFB were used to encrypt voice, resynchronization could be done at every connection attempt. For multiplexing systems, the start of every superframe or master group may be a convenient sync point.

For those systems which use Frequency Division Multiple Access (FDMA) techniques (i.e., most of the older military and civilian satellites and almost all long distance microwave systems) an appropriate technique is to increase the clock frequency of the signal being carried by just enough to accommodate an occasional synchronization word via bit stuffing techniques, e.g., adding an M-bit code word every N bits, and running the clock on the encrypted ("black") side at a rate (M+N)/N times the input clock rate. At the higher speeds, the conventional Bell System bit transparent speed of 1.344 Mbps may be used to carry the encrypted traffic, with the synchronization bits being inserted in the clear to bring the data rate up to 1.544 Mbps and still guarantee that the Bell System's constraint on the number of consecutive zeroes is met. Alternatively, the data plus the synchronization bits could all be encrypted, and the loss of synchronization detected after decryption based upon the absence of the framing bits. In this case, however, it is somewhat more difficult to differentiate between loss of crypto synch and the complete loss of the communications link. The two techniques, therefore, both have their advantages and disadvantages, and the choice between them must be made on the basis of systems considerations beyond the scope of this paper.

For the newer Time Division Multiple Access (TDMA) satellite systems, highly accurate timing is provided as an integral part of the overall system, and cryptographic synchronization is essentially free so long as crypto is built into the earth station. If not, then the same techniques used for FDMA systems can still be used, except that the TDMA system will provide the clocking to the crypto unit, and the data terminal equipment may run at just a little bit less than the rated speed.

Resychronization, of course, requires a new IV, and this brings up the possibility of repeating some portion of the cipher stream. If high data rates and/or very frequent reinitializations are expected, there are a number of potential solutions to the transmission in depth (i.e., reuse of the cipher stream) problem.

The first is obvious: change the key(s) frequently, so that not as many reinitializations occur before the key is changed. According to the calculations in Appendix B, up to ten thousand random reinitializations (IV reseedings) would result in a probability of repetition of less than one in a million, but the probabilities begin to increase rapidly after that point. One hundred thousand reinitializations would be marginal, and a million would clearly be unacceptable. (These calculations assume a 64-bit IV, and on the order of 300,000 DES cycles between reinitializations. More cycles would make the probabilities worse, but fewer cycles would have relatively little effect.)

The second solution might seem to be reinitialize in a manner that is not random, but rather is guaranteed to be unique (such as an incrementing counter). Although this guarantees that two cycles will never be generated that begin at the same point, it is still possible that a portion of the cycle that began with the second IV would overlap the beginning, end, or middle of the first partial cycle. The longer the sequence, or the more reinitializations, the wrose this problem becomes. Although Appendix B does not cover this case, the numerical results should be similar to those presented above.

5.0 RECOMMENDED KEY CHANGE SCHEDULE

The following schedule of key changes is recommended for 64-bit OFB, based on the analysis in Appendix B:

1. Change data encryption keys (working keys) after 10,000 reinitializations, four billion DES cycles, or one day, whichever comes first. This implies changing keys once a day for T1 encryption units, or twice a day for T2 units.

2. Change key encrypting keys (master keys) after every 10,000 working key generations, if random or pseudo-random key generation techniques are used. If deterministic means are used, e.g. repetitive encryption a la Output Feedback, or encrypting an incrementing counter, then changing the master keys on the order of once a week to once a month is recommended on the basis of general prudence.

3. If an entire set of master keys are stored in a crypto unit that complies with the tamper resistance requirements of Federal Standard 1027, then change them at least every two months, again based on the principle of limiting losses as much as on cryptographic arguments. Replenishing master keys once a month on a rolling basis, advancing from master key to master key once a week is recommended if sufficient master key storage is available.

The above figures are only slightly conservative, and are based upon a one chance in a million probability of a repetition, assuming approximately 300,000 DES cycles or about twelve seconds between reinitializations at the T1 rate. Because the probability of a repetition grows with the square of the number of reinitializations, the 10,000 limit is the most significant, and perhaps the least conservative. If reinitialization should occur every eight seconds on the average, key changes should be scheduled twice per day. For T2 rates, the limitation of four billion cycles is the controlling factor.

The rate at which master keys are changed should also be influenced by the perceived probability of a human compromise occurring. Based on the ABSCAM investigations, it could be argued that there is approximately one chance in a hundred that a randomly selected, presumably responsible person having access to the master key or keys could be compromised (either accidentally, or for a specified amount of money or other consideration). If so, then the possibility that a given set of keys will be compromised is $1 - .99^N$, where N is the number of people who have access to the keys. If multiple crypto units are netted together using a common key, then the probability of a system compromise grows rather rapidly. If 25 people have access to a single set of keys, the probability of a compromise under those circumstances would be an unacceptably high 22 percent! Although changing keys often may not change human nature, it does tend to increase the risk of exposure in the case of a deliberate compromise, and it, therefore, increases the cost of the attack. It also reduces the amount of information lost per incident, although that may be of little comfort.

The solution to the problem of compromise of the master keys lies in using two (or more) couriers per crypto unit, with each carrying what Blakley would call a site specific "instance" of the network master key or else a "shadow" of that same key (the terms are interchangeable). By recombining the two partial keys inside the crypto unit to form the true common network key, collusion between the two couriers at a single site would be required to compromise the system. An example of such a system would be the classical one-time pad, where the exclusive OR for decryption would take place inside the crypto unit. For multiple sites, multiple pairs of keys would be formed which when XORed together would produce the common master key, but separately they would be nothing but pure random numbers providing Shannon-perfect security. See Blakley[13] and Shamir[14] for further details and extensions.

6.0 SUMMARY

The author's recommendation would be to use the DES Output Feedback mode whenever a non-error propagating mode is required by system design considerations, but with the proviso that only full block (K=64 bit) feedback be implemented, and that consideration be given to the physical media involved to ascertain whether protection against the active threat is required. The cryptoperiod of the data encrypting (working) keys should be limited to 10,000 reinitializations, four billion DES operations, or one day, whichever comes first. Master keys should be changed once a week to once a month. The problems of networks operating with common encryption keys must be considered carefully to ensure that the risk of human compromise is not unacceptedly high.

7.0 REFERENCES

1. DES modes of operation, Federal information processing standards publication 81. National Bureau of Standards, Sept. 25, 1980.
2. Data encryption standard, FIPS PUB 46. U.S. Dept. of Commerce/National Bureau of Standards, Jan. 15, 1977.
3. Orceyre, M. J., and Heller, R. M. An approach to secure voice communication based on the data encryption standard. IEEE Communications, Nov. 1978, pp. 41-50.
4. Campbell, C. M. Design and specification of cryptographic capabilities. IEEE Communications, Nov. 1978, pp. 15-19.
5. Davies, D. W. and Parkin, G.I.P. The average cycle size of the key stream in Output Feedback encipherment. Advances in Cryptography: Proceedings of CRYPTO 82. Plenum Publishing Corp., 233 Spring Street, New York, NY 10013.
6. Proposed Federal Standard 1025. Telecommunications: Interoperability and security requirements for the use of Data Encryption Standard in the network and transport layers of data communications. National Communications System, Washington, D.C. 20305. Draft of June 1, 1981.
7. Proposed Federal Standard 1026. Telecommuncations: Interoperability and security requirements for the use of Data Encryption Standard in the physical and data link layers of data communications. National Communications System, Washington, D.C. 20305. Draft of June 1, 1981.
8. Diffie, W., and Hellman, M. E. Privacy and authentication: an introduction to cryptography. Proceedings of the IEEE, Vol. 67, No. 3, March 1979, pp. 397-427.
9. Knuth, D. E., The Art of Computer Programming; Volume 1: Fundamental Algorithms. Reading, MA: Addison Wesley.
10. Knuth, D. E., The Art of Computer Programming; Volume 2: Seminumerical Algorithms. Reading, MA: Addison Wesley.
11. Gait, J. A new non-linear pseudo-random number generator, IEEE Transactions on Software Engineering. Vol. SE-3, No. 5, Sept. 1977, pp. 359-363.
12. Hellman, M. E. and Reyneri, J. M. The distribution of drainage and the DES. Advances in Cryptography; Proceedings of CRYPTO 82. Plenum Publishing Corp., 233 Spring Street, New York, NY 10013.
13. Blakley, G. R. Safeguarding cryptographic keys. Proceedings of the National Computer Conference, 1979. AFIPS Press, Vol. 47 (1979), pp. 313-317.
14. Shamir, A. How to share a secret. Comm. of the ACM, Vol. 22 (1979), pp. 612-613.

15. Meyer, C. H. Ciphertext/plaintext and ciphertext/key dependence vs. number of rounds for the data encryption standard. Proceedings of the 1978 National Computer conference, AFIPS Press, Montvale, NJ.
16. Davies, D. W. Private communication, August 3, 1982.

APPENDIX A. WEAK AND SEMI-WEAK KEYS

Because the Output Feedback mode is data independent, it would be particularly susceptible to a "weak" key, i.e., a key that would cause a shorter than average cycle to be generated. Therefore, let us consider what causes the four known weak keys to be weak.

The four known weak keys are (in hexadecimal):

```
01 01 01 01 01 01 01 01
FE FE FE FE FE FE FE FE
1F 1F 1F 1F 0F 0F 0F 0F
E0 E0 E0 E0 F1 F1 F1 F1
```

Referring to the definition of DES in FIPS PUB 46[2], it will be noticed that the 56-bit-plus-parity (64-bits in total) key is used to generate sixteen 48-bit keys, one of which is used in each of the two parts, C(0) and D(0). The bits making up the C(0) part are derived from the original key in accordance with the first 28 positions of the Permuted Choice 1 table:

```
57 49 41 33 25 17  9 1
58 50 42 34 26 18 10 2
59 51 43 35 27 19 11 3
60 52 44 36
```

The D(0) is derived similarly, per:

```
63 55 47 39 31 23 15 7
62 54 46 38 30 22 14 6
61 53 45 37 29 21 13 5
28 20 12  4
```

Once the C(0) and D(0) blocks are obtained, the C(i) and D(i) blocks are derived by a series of circular left shift operations, by the following schedule:

1. Iterations numbered 1, 2, 9, and 16 are derived from the previous iteration by a one-bit circular left shift.

2. All other iterations use a 2-bit circular left shift.

The 48-bit key for round N, K(N), is obtained by selecting bits from the concatenation of C(N) and D(N) in accordance with Permuted Choice 2, ignoring bits 9, 16, 18, 22, 25, and 35:

```
14 17 11 24  1  5
 3 28 16  6 21 10
23 19 12  4 26  8
16  7 27 20 13  2
41 52 31 37 47 55
30 40 51 45 33 48
44 49 39 56 34 53
46 42 50 36 29 32
```

It may now be apparent why the weak keys are in fact weak - their C(0) and/or D(0) blocks are either all zeroes or all ones. In either case, the results of the left shifts will be the same for all sixteen iterations. Since decryption is exactly the same as encryption except for the fact that the key schedule and rounds are applied in the reverse order, the weak keys generate an ecrypt, decrypt, encrypt, decrypt... sequence. The result of double ecnryption using a weak key twice is therefore the same as no encryption at all!

According to Meyer[15], and for the above reason, five rounds of the basic DES kernel are necessary to achieve complete intersymbol dependence between ciphertext and plaintext, or between ciphertext and key. If the key were such as to generate the 01's or 10's pattern in their C(0) and/or D(0) blocks, only the bare minimum five rounds would have significant key variation and therefore a limited amount of interdependence upon all of the bits of the key and the input, although all of the other attributes of DES such as the S-boxes would continue to operate as usual. Although a limited experiment with one such key did not cause the key stream to repeat within a billion iterations, it would probably be prudent to avoid these 12 semi-weak keys.

In addition, the pattern 00110011001100110011001100110011 (and its complement), although stronger than the alternating ones and zeroes pattern, generates only four unique keys in the key expansion schedule. As a matter of fact, as Davies[16] pointed out, any repeating four bit pattern in the C(0) or D(0) block will generate only four unique subkeys during the 16 rounds, as follows:

```
C(0)     ABCD ABCD ABCD ABCD ABCD ABCD ABCD
C(1)     BCDA BCDA BCDA BCDA BCDA BCDA BCDA
C(2)     CDAB CDAB CDAB CDAB CDAB CDAB CDAB
C(3)     ABCD ABCD ABCD ABCD ABCD ABCD ABCD
C(4)     CDAB CDAB CDAB CDAB CDAB CDAB CDAB
C(5)     ABCD ABCD ABCD ABCD ABCD ABCD ABCD
C(6)     CDAB CDAB CDAB CDAB CDAB CDAB CDAB
C(7)     ABCD ABCD ABCD ABCD ABCD ABCD ABCD
C(8)     CDAB CDAB CDAB CDAB CDAB CDAB CDAB
C(9)     DABC DABC DABC DABC DABC DABC DABC
C(10)    BCDA BCDA BCDA BCDA BCDA BCDA BCDA
C(11)    DABC DABC DABC DABC DABC DABC DABC
C(12)    BCDA BCDA BCDA BCDA BCDA BCDA BCDA
C(13)    DABC DABC DABC DABC DABC DABC DABC
C(14)    BCDA BCDA BCDA BCDA BCDA BCDA BCDA
C(15)    DABC DABC DABC DABC DABC DABC DABC
C(16)    ABCD ABCD ABCD ABCD ABCD ABCD ABCD
```

Since there are 16 possible patterns of four bits that might be entered into the C(0) register, and the same number for the D(0) register, there are 256 different DES keys of this type. However, we should exclude the four weak keys and the 12 semi-weak keys, so there are a total of 240 keys we will call "demi-semi-weak".

(It should be said that this regularity does not appear to cause any particular weakness. It only serves to highlight some areas for further investigation.)

Finally, keys containing arbitrary bit patterns in their C(0) or D(0) blocks together with the all zeroes, all ones, 01...01, 10...10, and/or ABCD...ABCD pattern in the other block will be called "hemi-demi-semi-weak" (with tongue in check and apologies to cryptologists who are also musicologists). They should be avoided if it is convenient to screen them automatically.

The weak and semi-weak keys are listed for reference along with the 20 demi-semi-weak keys that involve the 0011...0011 or 1100...1100 pattern in either the C(0) or D(0) register, or both.

The moral is obvious: BEWARE OF COURIERS BEARING KEYS MADE UP OF 0'S, 1'S, E'S, AND F'S PREDOMINATELY. If keys are created using strictly random processes, these pathological cases will be a highly unlikely occurrence. It should also be mentioned that these or other nefarious possibilities could potentially be created by a Trojan horse key generator installed by an attacker, or by possible failure mechanisms. It follows that any hardware (and/or software) used to generate keys automatically should be subjected to the same rigorous certificational processes and self-tests that are recommended for the DES algorithm itself.

Table A-1. Weak, Semi-Weak, and Demi-Semi-Weak Keys

```
WEAK KEYS:                          DEMI-SEMI-WEAK KEYS:
    0101 0101 0101 0101                 E0E0 0101 F1F1 0101
    FEFE FEFE FEFE FEFE                 FEFE 1F1F FEFE 0E0E
    1F1F 1F1F 0E0E 0E0E                 0101 E0E0 0101 F1F1
    E0E0 E0E0 F1F1 F1F1                 1F1F FEFE 0E0E FEFE
                                        1F1F 0101 0E0E 0101
SEMI-WEAK KEYS:                         FEFE E0E0 FEFE F1F1
    E001 E001 F101 F101                 0101 1F1F 0101 0E0E
    FE1F FE1F FE0E FE0E                 E0E0 FEFE F1F1 FEFE
    01E0 01E0 01F1 01F1                 FE1F E001 FE0E F101
    1FFE 1FFE 0EFE 0EFE                 E001 FE1F F101 FE0E
    1F01 1F01 0E01 0E01                 1FFE 01E0 0EFE 01F1
    FEE0 FEE0 FEF1 FEF1                 01E0 1FFE 01F1 0EFE
    011F 011F 010E 010E                 FEE0 1F01 FEF1 0E01
    E0FE E0FE F1FE F1FE                 E0FE 011F F1FE 010E
    FE01 FE01 FE01 FE01                 FEFE 0101 FEFE 0101
    E01F E01F F10E F10E                 E0E0 1F1F F1F1 0E0E
    1FE0 1FE0 0EF1 0EF1                 1F01 FEE0 0E01 FEF1
    01FE 01FE 01FE 01FE                 011F E0FE 010E F1FE
                                        1F1F E0E0 0E0E F1F1
                                        0101 FEFE 0101 FEFE
                                        (Partial list).....
```

APPENDIX B. ANALYSIS OF OFB CRYPTOPERIOD

B.1 Analysis of OFB Without Reseeding

Consider a cycle of 64-bit block outputs produced by OFB from an initial seed value. We know that because the DES produces a unique output from a given input and vice versa, the cycle produced from that seed value has no points in common with any other cycle that could be produced with that key. We do not as yet know how to compute the length of any given cycle, nor how many cycles might be produced by a given key, but if we had access to a sufficiently powerful DES engine, we could simply run it until we answered those questions.

If we had such a machine, we could start it operating at some random point and let it continue until the cycle closed upon itself. Then we could restart the process, taking care to use a new seed value that was not contained in the first cycle. That process would run until it finally closed upon itself, ending the second cycle, and so forth. (Not only must we have a very powerful DES engine, but we must have a 2^{64} bit memory as well, so that we could keep track of what variables had already been visited by this cryptographic version of a Knight's Tour.)

If we were to continue this hypothetical experiment until the entire space of 2^{64} variables had been exhausted, we could rank the cycles in order by their length, with the shortest cycle (of length J1) coming first. If we had the results of such a ranking the next time we randomly chose a new starting value, we would at least know FOR THAT GIVEN KEY there was a guaranteed minimum cycle length (J1), and that depending upon which cycle were selected by virtue of the initial seed value, the probability of repeating the seed value would be zero for I<J iterations, and would increase to some non-zero value some time after that point.

If we then passed the I=J1 point without repeating the initial seed value, we would know that the cycle we were then (re-)exploring was not the shortest cycle, and that we could therefore eliminate from the space of $M=2^{64}$ possible points not only the I=J1 points we had already tested as part of this cycle, but the J1 points that were contained in that shortest cycle.

Eventually, we might also pass the I=J2 point in our iterations, where J2 is the length of the second shortest cycle for this key. We would then have ruled out J1 + J2 + I variables from the search space, even though we had only explored I=J2 possibilities while looking for the end of the particular cycle determined by the initial seed. Continuing this process, we would have exhausted a superincreasing sequence of possibilities J1 + J2 +...+ Jn + I after exploring only I=Jn variables. The difficulty is that we cannot really run the hypothetical process to determine all the Jn's, so we do not know how much progress we are making, or what the size is of the variable space that remains after N iterations. However, we may be able to obtain an estimate of the cycle length through the following, subject to experimental confirmation.

The design of the DES algorithm plus the available evidence suggests that DES in OFB mode should be a good random number generator. We will therefore make the assumption that the output of K=64 OFB after every iteration can be approximated by a random-draw-without-replacement process, where the initial seed value has been cast back into the space of 2^{64} variables. We will also temporarily neglect all cycles other than the shortest one, to which we will assign a length J.

Then the probability that the first output is the same as the seed would either be zero, or 1/M; depending upon whether J=1. The probability that the second output is the same as the seed would then be either zero, or 1/M(-1). In general, the probability that the I-th output is the same as the seed would be 0 for I<J, and 1/(M-I+1) for I≥J. The probability that the I-th output is NOT the seed is, therefore, 1 for I<J, and 1 - 1/(M-I+1) or (M - I)/(M - I + 1) for I≥J.

The probability that none of the output values have yet equalled the seed after N=>J cycles is, therefore, the product of all of these probabilities. Observing that the I-th numerator and the I-1st denominators can be cancelled, we have

$$P = \prod_{I=J}^{N} \frac{M-I}{M-I+1} = \frac{M-N}{M-J+1}.$$

The probability that the cycle has begun to repeat is therefore

$$\text{Prob(repeat)} = 1 - P = \frac{N-J+1}{M-J+1}.$$

Now the question is how to treat the minimum cycle length J, which we have no way of knowing. If J is in reality fairly large, then for small values of N<J the probability of a repetition will really be zero, and we are safe no matter what we assume. On the other hand, if N is small compared to M, and if we assume that there may be a small number of very short cycles such that J approaches 1, then we may tend to overestimate the probability of repetition, which is at least conservative.

As a result, we can estimate the probability that there will be no repetition of the initial seed value after N iterations by:

$$\text{Prob(repeat)} = 1 - P = N/M = \frac{N}{2^{64}}.$$

Davies[5] stated essentially the same result, and in a letter[16] offered several other intriguing observations. These hints forced the author to go back to Knuth[9,10], who treats several of these problems in the text and the exercises in a much more rigorous and elegant manner than the above. Assuming that 64-bit OFB can be modelled as a random set of permutations of 2^{64} objects (selected by the key) then according to Knuth:

1. The average number of OFB cycles of length J is 1/J, with a deviation equal to the square root of 1/J. (Vol. 1, section 1.3.3, exercise 18.)

2. The probability that a key will produce one or more cycles of length one (i.e., the identity encryption/decryption operation) is $1 - (M-1)^M/M^M$, which converges to $1 - 1/e$, or 63%. (Vol. 2, section 3.1, exercise 15.)

3. The probability of a repetition should be N/M + 1/2. (Vol. 1, section 1.3.3, exercise 17.)

4. The average length of the longest cycle in a permutation of degree M is approximately $0.62433^{(M+\frac{1}{2})}$, where M in this case is 2^{64}. (Vol. 1, section 1.3.3, exercise 23.)

This confirms that the average cycle length should be approximately 2^{63}, assuming that the random draw model is valid. Although we think that it is probably true, the effort to validate such a hypothesis is equivalent to an exhaustive search squared, and so we will probably never know.

To be conservative, we should probably limit the number of OFB cycles without reinitialization or key change to substantially less than 2^{63}, both to limit the possibility of a transmission in depth occurring because of an error in our assumptions and to limit our losses if such a compromise did occur. Because the detection of the occurrence of a cycle would be relatively easy if OFB were never reinitialized (because the data could be analyzed in real time as the ciphertext is intercepted, and would require the storage of only a relatively few blocks of data), the probability of a cycle occurring should be kept quite low in this case. This question will be addressed again at the conclusion of this appendix.

B.2 Analysis of OFB with Random Reseeding

Another interesting question is what the probable cycle length would be if the OFB cycle were occasionally interrupted and randomly reseeded, for example to restart or resynchronize a communications link periodically, or in an attempt to avoid the possibility of short cycles by sampling another part of the space. In this case, we will be on much firmer ground with respect to the assumptions.

If a series of random values were input to the DES algorithm, we could estimate the probability that NOT EVEN A SINGLE PAIR of outputs would have been the same after N outputs by solving the so-called birthday problem, which computes the probability that no two people out of the N people in a room will have the same birthday.

Suppose that a genuinely random value were loaded as a DES input, and a single 64-bit output value produced, and then a second output were produced by encrypting a second random value. The probability that the second output is not equal to the first is $(M - 1)/M$, where $M = 2^{64}$. In general, the probability that the i-th value is not equal to any of the previous i-1 values is $(M - i)/M$. The probability that all of the values are different from each other is the product

$$P(\text{no repeat}) = \prod_{i=1}^{N} \frac{M-i}{M} = \frac{M!/(M-N)!}{M^N}.$$

This formula can be evaluated using the natural log form of Sterling's approximation for the factorial function, if the calculations are performed in extended precision (33 digits) to retain the required accuracy. The log of the probability of no repeat is given by

LOG_P = STERLING(M) - STERLING(M-N) - N*LOG(M),

which can be expanded to

LOG_P = (M+1/2)*LOG(M)-M+LOG(SQRT(2*PI)) -
        ((M-N+1/2)*LOG(M-N)-(M-N)+LOG(SQRT(2*PI))
        - N*LOG(M).

This can be simplified by combining terms (which also helps preserve the necessary precision), to produce

P = EXP(LOG_P) = EXP((M-N+.5)*(LOG(M)-LOG(M-N)) - N).

From this, it can be calculated that the probability of no repetition is about 95% after 1.34 billion probes of length one, and about 50% after 5.0 billion probes; in comparison to the undisturbed OFB which would have a calculated probability of repetition of 2.7×10^{-10} after 5 billion probes.

As a rule of thumb which is accurate enough for many calculations, the probability of an overlap can be approximated by $P = N^2/2M$ for N up the square root of M.

B.3 Analysis of OFB in the General Case

The next question is what is the probability of any DES output being repeated if the IV is reseeded after X outputs, with this process being repeated Y times for a total of N = X*Y outputs.

We know that the probability of not repeating a value within a chain of length X is P = (M - X)/(M-J+1), where $M = 2^{64}$ and J is the presumed minimum guaranteed cycle length (or zero). The probability of not repeating within Y chains of length X is therefore $P = ((M-X)/(M-N))^Y$. The remaining problem is to estimate the probability of repetition or overlap between chains.

Assume that one chain of length X has been produced, and that now a second one is generated. The probability that the second chain will begin at such a point that it will overlap the first is

(X-1+X)/M. The probability that it will not overlap is, therefore, (M-X+1-X)/M. The probability that the i-th chain will not overlap any of the previous chains is, therefore, ((M-X+1)-i*X)/M, and the probability that none of the chains will overlap between themselves is, therefore

$$P(\text{no overlap}) = \prod_{i=1}^{Y} \frac{((M-X+1)/X)-i}{M/X}, \quad \text{or} \quad \prod_{i=1}^{Y} \frac{Q-i}{M/X}.$$

where $Q = (M-X+1)/X$. This is equivalent to

$$P(\text{no overlap}) = \frac{Q!/(Q-Y)!}{(M/X)^Y}.$$

It will be noticed that if X is one, the probability derived from the birthday problem is obtained.

The general form of the probability that no overlap between full 64-bit values will occur, either within or between the Y strings, is therefore

$$P = \frac{(Q)!/(Q-Y)!}{(M/X)^Y} * \frac{(M-X)^Y}{(M-J+1)^Y}$$

Applying Sterling's approximation in natural log form again, we have

$$\text{LOG\_P} = \text{STERLING}((Q) - \text{STERLING}((Q-Y) - Y*\text{LOG}(M/X) + Y*\text{LOG}(M-X) - Y*\text{LOG}(M-J+1).$$

From the previous results, we have

$$\text{LOG\_P} = (Q-Y+.5)*(\text{LOG}(Q)-\text{LOG}(Q-Y) -Y- Y*(\text{LOG}(M/X) + Y*(\text{LOG}(M-X) - \text{LOG}(M-J+1)).$$

The following tables present the probability of repeating any portion of the keystream output presented by OFB at the rate of one 64-bit output per microsecond, assuming the IV is randomly reseeded once every 300 milliseconds, once a day, or once every 100 days, and again assuming N is less than the square root of N and that J approaches 1.

The quantity Y indicates the multiple of the reseeding interval.

Table B-1. Probability of Repetition After 300 MS Reseed

X = 300,000   (One per microsecond for Y * 300 ms)

Y	P(REPEAT)	P(NO REPEAT)
1.0000000E+02	5.96046E-08	9.99999E-01
1.0000000E+03	5.96046E-08	9.99999E-01
1.0000000E+04	8.34465E-07	9.99999E-01
1.0000000E+05	8.13603E-05	9.99918E-01
1.0000000E+06	8.09860E-03	9.91901E-01
2.0000000E+06	3.20028E-02	9.67997E-01
4.0000000E+06	1.21996E-01	8.78003E-01
4.8000000E+06	1.70846E-01	8.29153E-01
5.7600000E+06	2.36455E-01	7.63544E-01
8.2944000E+06	4.28462E-01	5.71537E-01
1.7199267E+07	9.09772E-01	9.02272E-02

Table B-2. Probability of Repetition After Daily Reseed

X = 8.64000E+10   (One every 1.0 microseconds for Y days)

Y	P(REPEAT)	P(NO REPEAT)
1	5.96046E-08	9.99999E-01
2	5.96046E-08	9.99999E-01
4	1.19209E-07	9.99999E-01
8	2.38418E-07	9.99999E-01
16	7.15255E-07	9.99999E-01
32	2.62260E-06	9.99997E-01
64	1.00731E-05	9.99989E-01
128	3.92794E-05	9.99960E-01
256	1.55270E-04	9.99844E-01
512	6.17325E-04	9.99382E-01
1024	2.45982E-03	9.97540E-01
2048	9.78875E-03	9.90211E-01
4096	3.85562E-02	9.61443E-01
8192	1.45484E-01	8.54515E-01
16384	4.66754E-01	5.33245E-01
32768	9.19131E-01	8.08685E-02

Table B-3. Probability of Repetition After 100 Day Reseed

X = 8.64000E+12   (One every 1.0 microseconds for Y * 100 days)

(N.B.: The use of such large numbers of iterations may not be a valid use of the model, and are not recommended.)

Y	P(REPEAT)	P(NO REPEAT)
1	4.76837E-07	9.99999E-01
2	1.90734E-06	9.99998E-01
4	6.13927E-06	9.99993E-01
8	2.01463E-05	9.99979E-01
16	7.07507E-05	9.99929E-01
32	2.61843E-04	9.99738E-01
64	1.00326E-03	9.98996E-01
128	3.91882E-03	9.96081E-01
256	1.54078E-02	9.84592E-01
512	5.98868E-02	9.40113E-01

Finally, we must consider what degree of risk we can withstand with respect to a transmission in depth exposure.

To be safe, let us assume that the value of the information to be protected by OFB is worth a maximum of $1 billion in the aggregate, but that no single piece of information would be worth that much. (An example might be a major energy discovery, similar in scope to the Alaskan off-shore oil field discovery. To obtain the full value of the information, it might be necessary to obtain the precise locations of every test hole and the full details of the seismology, core samples, productions, etc. But the fact that a major discovery had been made somewhere off-shore of Alaska might be sufficient for a competitor to start drilling, trying to acquire leases, etc.)

Let us therefore speculate that an attacker would have to break a solid month's worth of traffic to gain all of that value, but that the old 80/20 rule of thumb applies here also, i.e., 80% of the value could be obtained from 20% of the traffic, or $800,000,000 in 6 days, or $266,666,666 per day on the average.

Assuming that the traffic is being carried via a 1.544 Mbps T1 link, or approximately one encryption operation every 41.45 microseconds, then 2^{31} iterations (1.03) days would result in a probability of $2^{31}/2^{64} = 1.164*10^{-10}$ of a cycle occurring. At least an additional day's worth of traffic would then have to be obtained to recover the first day's traffic and all subsequent information until the key is changed. In the zero sum game sense, the attacker's expectation of profit is $266,666,666*1.164*10^{-10}$, or $0.0031 per day, and even a year would only net him $1.13.

On the other hand, if the probability of repetition were to rise to 10 per million, then the attacker's expectation would be $97,333 at the end of the year. That is enough to lease a reasonably powerful mini-computer system, and with a few lucky breaks that just might be enough power to break some portion of the traffic. For that reason, it is suggested that the probability of even a single repetition occurring be held to under one chance per million. (If the traffic volume is significantly less than the one microsecond rate, it is conceivable that even a micro-processor system could be used to attack the traffic. But if the working keys are changed once per day regardless of the traffic rate, there will be correspondingly fewer iterations and a lower probability of success.)

Tables B-1 through B-3 indicate the risk of compromise in the more general case, so users can see the threat that they may be exposed to. The rate of one microsecond per encryption operation is assumed, so as to cover some of the faster applications that may be coming.

Based upon that data, it would appear that it MIGHT be safe to run OFB in a continuous output mode for about 50 minutes before changing keys if the IV is reseeded every 300 milliseconds (10,000 reinitializations), 32 days if the IV is reseeded once a day, or about one year if the IV is reseeded every 100 days; assuming the one microsecond encryption rate, and assuming that a one chance in a million probability of a repetition is satisfactory.

However, the author has less than perfect confidence in some of the foregoing assumptions. In particular, some keys may generate distributions of cycles which are very far from the average, and yet such a key might be the one that was in use when a particularly valuable piece of information was transmitted. In addition, very long strings of OFB output without reinitialization would be the easiest type of traffic to analyze for a transmission in depth, and would yield the most information as well.

Table B-2 would suggest that if OFB is reseeded or reinitialized daily, then up to 16 days between key changes would not lead to an excessive probability of repetition at the microsecond rate. But even with the reinitializations that is a very substantial amount of traffic, probably more than should be risked by encryption under any given key. Considering the random reinitialization analysis as a conservative worst case, it would be desirable to limit N to 2^{32}, or approximately four billion operations or 256 gigabits of traffic. It is, therefore, recommended that working keys be changed after 4 billion encryption operations, 10,000 reinitializations, or one day, whichever comes first.

Finally, if some random or psuedo-random process is being used to generate new working keys, then the birthday problem analysis applies to the generation of duplicate working keys. Although the occurrence of a duplicate working key might be difficult to detect, the effect would be catastrophic. The probability of such a duplication should be held to well below one chance in a billion, or $P = N^2/2^{57} = 10^{-9}$. N should, therefore, be held to less than 10,000. If the working keys are generated by repetitively encrypting the working key under control of a master key in a manner similar to OFB, then $P = N/2^{56} = 10^{-9}$, so N should be less than 72 million. Even if working keys were generated at the rate of once every 300 milliseconds, this would amount to a key change being required every 7.6 years, and of course it would be advisable to change even master keys much more often than that on grounds of general prudence.

DRAINAGE AND THE DES
SUMMARY

Martin E. Hellman and Justin M. Reyneri

Information Systems Laboratory
Stanford University
Stanford, California 94305

Introduction

In our paper, we investigate a statistical property of random functions we named drainage. (Definitions for drainage, random function, etc. will be given shortly.) Our motivation for doing so is twofold. First, it generally assumed that a good cryptographic system will exhibit no simple statistical regularity. For example, the function from key to ciphertext when a block cipher is used to encode a fixed plaintext should appear to be completely random. We were therefore interested in studying the behavior of drainage for a random function, and then comparing it to the measured behavior for a real cryptosystem, the DES. Secondly, drainage is closely related to statistical properties which are important to the performance of the generalized cryptanalytic attack proposed by Hellman [1], discussed below.

The random functions we deal with are mappings from finite sets into themselves. For our purposes, these functions can be thought of a being implemented by a lookup table in which each entry is chosen uniformly, independently and with replacement from the potential elements in the range. We sometimes find it convenient to think in terms of the directed graph that is naturally associated with each realization of a random function, i.e. the graph having a vertex for each element of the domain of the function, and a directed edge from vertex x to vertex y iff $y = f(x)$. Such graphs consist of a number of components, each of which has exactly one cycle. If an arbitrary vertex in the graph is chosen and the path beginning there is followed, it will eventually end in such a cycle. (This is just another way of viewing the fact that iteration of a function from a finite set into itself eventually leads to periodic repetition of the same values; the fact that each component of the graph has only one cycle is equivalent to the fact that the function is single-valued.) In drawing these graphs, the cycles appear to be lakes and the paths leading into them look like tributary systems. We therefore refer to the number of vertices in a component as the drainage into that component.

In Hellman's cryptanalytic method [1], a block cipher is used as a function from the key space into itself by fixing the plaintext at a chosen value and interpreting the output of the system, the ciphertext, as an element of the key space. In the case of DES, where keys are 56 bits long and ciphertext blocks are 64 bits long, a simple reduction function is used to obtain a 56 bit value from the ciphertext. (In our experimental work, we simply discard the parity bit of each byte of ciphertext.) The structure of the random graph that results when a cryptographic system is operated in this way is important to the behavior of the attack. A highly favorable structure would be on in which the graph consisted only of cycles, with no "rivers" draining into them. In such a case, the method could break a system in $N^{1/2}$ operations using $N^{1/2}$ words of memory, where N is the number of elements in the key space. Such a favorable structure is highly improbable. Under very plausible assumptions about the structure of a the graphs obtained using his attack, Hellman shows that it should be effective in $N^{2/3}$ operations and the same amount of memory. On the other hand, he points out that there are some unlikely structures against which the attack would be ineffective. All of the above estimates refer to a postcomputation which can only be performed after a precomputation which is equivalent to exhaustive search. The method is of interest because the precomputation is easily done in parallel, and can be reused to obtain an arbitrary number of keys.

The drainage into the components of a random graph is not directly related to the performance of Hellman's approach. We looked at drainage and the DES as an initial attempt to determine the typicality of the graphs produced using DES.

Theoretical Results

We obtained an asymptotic expression for the probability density for the drainage of the largest component of a random graph in a straightforward way. To give the flavor of the approach, we consider the case where this drainage accounts for more than half of the points in the graph. In this case, the probability that the largest component in a graph on N points has drainage exactly equal to i is given by

$$P_N(i) = (1/N)^N \binom{N}{i}(i/N)^i((N-i)/N)^{(N-i)}C(i) \tag{1}$$

where $C(i)$ is the probability that a random graph containing i points consists of a single component. The first term above is just the number of directed graphs on N points; the second is the number of ways to choose i points for the component with i points. The next two terms reflect the constraints that the subset of i points map only into themselves while the remaining points map into themselves. Finally, there is the requirement that i-point subset be a single component. In a way which is simpler than others we know of, we rederived the asymptotic expression for $C(i)$, that is, $C(i) = \sqrt{\pi/i}$. Using this expression and Stirling's formula, (1) simplifies to

$$P_N(i) = (k/i)\sqrt{N/(N-i)}. \tag{2}$$

This density is roughly constant for values of drainage between 50% and 90% of N, and increases significantly for drainage of 90% or more.

The probability density for drainage less than 50% was obtained in a similar fashion. Using it, we estimate the expected value of the drainage into the largest component to be near 80%; it also proved useful in explaining some surprising experimental results.

Experimental Results

We performed two related experiments to estimate the drainage in graphs obtained from DES. In both, we took advantage of the fact that certain aspects of the structure of a directed graph (or the equivalent function) can be studied even though only a negligible fraction of the points in the graph are examined. (This is necessary because a complete study of the graph is at least as difficult as exhaustive search through the DES key space.) In particular, if we begin at an arbitrary point in the graph and follow the path leading from it, we will almost certainly enter a cycle before traveling farther than a few multiples of $\sqrt{N}$. In addition, it is possible to detect the fact that we cycled and to compute the cycle length with little additional effort. By choosing different starting points and calculating the lengths of the cycles into which they drain, we can tell when different points drain into the same component. Thus, we can estimate the size of the largest component in a random graph on N points with only $N^{1/2}$ operations.

In our first experiment, we fixed the plaintext at the all-zero value and chose 100 different starting values at random. We were surprised to find that they all drained into the same component, for this led to an estimate of at least 99% drainage into the largest component, significantly larger than we had expected. The explanation lies in the relatively high values for the probability density high values of drainage. While the expected value of drainage is near 80%, the chances of it being near 100% are significant. By always using the same fixed plaintext, we were always examining the structure of the same graph; it was necessary to change the plaintext value used in the experments in order to investigate other graphs in the family generated by DES. In our second experiment we used a variety of values for the fixed plaintext but used each value with only two starting points. By counting how many times a pair of starting points drain into the same component, we can estimate the average size of the largest component for all DES generated graphs. The results of experiments with 70 pairs of starting points (i.e. 70 different DES graphs), are in agreement with the predictions of out theoretical work.

Conclusion

We have been able to characterize the behavior drainage in a random graph. We have also taken advantage of the fact that it is easy to find cycles in a directed graph on a large number of points in a relatively short amount of time. Doing so allows us to compare the behavior of DES to a truly random function. The results of our experiments show no statistical irregularity in DES. We did not discover anything in our experimental work which has immediate impact on the cryptanalytic attack proposed by Hellman.

REFERENCES

[1] M. E. Hellman, "A cryptanalytic time-memory tradeoff," *IEEE Trans. Inform. Theory*, vol. IT-26, pp. 401-406, July 1980.

SECURITY OF A KEYSTREAM CIPHER

WITH SECRET INITIAL VALUE

Robert S. Winternitz

Stanford University
Stanford, California

ABSTRACT

A keystream can be produced by driving a known pseudo-random function with a counter, starting with a secret initial value. Knowledge of M blocks of keystream allows a speed-up of cryptanalysis by a factor of M over exhaustive search. A similar result holds when the function is a secret choice from a parameterized family. These results are the best possible under a black-box model, i.e., where the function is revealed to the analyst only by calling an oracle.

A synchronous cryptosystem can be produced by driving a pseudo-random function with a counter to generate a keystream. The initial value of the counter may be kept secret as part or all of the key. This paper shows that this provides some additional security, but not as much as might appear at first glance. The author wishes to thank H. Amirazizi, M. Hellman, E. Karnin, and J. Reyneri for useful conversations.

Naturally such a system depends upon the cryptographic security of the basic function. It must be one-way in a strong sense: given the values of the function on some set of arguments, it must be infeasible to compute the value on any argument not in the set.

For this paper, we assume that the basic function is a cryptographically secure pseudo-random function. We consider only attacks which are uniform, attacks which do not use particular weaknesses of the function. We present such an attack which reduces the

---

*Supported by NSF grant ECS79-16161

cryptanalyst's work by a factor of M over exhaustive search, when he is given M corresponding blocks of ciphertext and plaintext. Furthermore, we show that this is optimal for such uniform attacks.

In the first system we consider, the only secret is the initial value of the counter.

System A

Let f be a publically known function from binary numbers of length n to numbers of length c. Pick a secret starting value X and define $KS(i)$, the ith block of keystream, by $KS(i) = f(X+i)$. (Addition mod-2^n). The ciphertext is the mod-2 sum of the plaintext and the keystream: $C(i) = P(i) \oplus KS(i)$.

In the next system, there is an additional secret parameter.

System B

Here f is a function of two variables, of length k and n respectively. Choose a secret K and a secret X, and define $KS(i) = f(K, X+i)$.

> Example: DES in counter mode is such a system with k=56, n=64, and c the number of bits from each block, which are actually used in the keystream.

It is necessary to place some conditions on f to allow the cryptanalyst to determine X and K. For instance, if f is a constant, then X cannot be determined regardless of the amount of computation or known keystream. Similarly, if f is almost always a constant, then extremely large amounts of keystream are needed to determine X. Of course, such functions are cryptographically very weak; although it is hard to determine X, the ciphertext can often be deciphered without knowing X.

Therefore, we assume that both K and X are always determined by some small amount of keystream; i.e., for some small b, whenever $f(K, X+1) = f(K', X'+1)$, $f(K, X+2) = f(K', X'+2)$, ... $f(K, X+b) = f(K', X'+b)$ then $K=K'$ and $X=X'$. For a random function we may take b around $2(k+n)/c$ with high probability. In any case, our algorithm will produce a pair K',X' such that the keystream it generates agrees with the real one on b blocks.

Theorem 1

If the analyst is given M consecutive blocks of keystream (cM bits), with $M \leq 2^{n/2}$ he can break system A in $O(2^n/M)$ steps, determining X (and hence the rest of the keystream). Given M blocks of keystream with $M \leq \min(2^{(n+k)/2}, 2^n)$, he can break system B in $O(2^{n+k}/M)$ steps.

Proof

Suppose the analyst is given M blocks of keystream $KS(1), \ldots, KS(M)$. He creates $M'=M-b+1$ strings, each consisting of b consecutive blocks, i.e., $KS(1) * KS(2) * \ldots * KS(b)$, $KS(2) * KS(3) * \ldots * KS(b+1)$, $\ldots$ $KS(M') * \ldots * KS(M)$, and stores them in a hash table. Thus this table contains $f(X+1) * \ldots * f(X+b)$, $f(X+2) * \ldots * f(X+b+1)$, $\ldots$ $f(X+M') * \ldots f(X+M)$, for the unknown X. This takes $O(M)$ steps and words of memory.

To break system A, he then evaluates f at b consecutive values, starting from each multiple of M', i.e., he creates the sequence $f(o) * f(1) * \ldots * f(b-1)$, $f(M') * \ldots * f(M'+b-1)$, $\ldots f(iM') * \ldots * F(iM'+b-1), \ldots$ for $iM' < 2^n$. He checks each such string against the hash table. Since there must be a multiple of M' between $X+1$ and $X+M'$, he must find a match; for some i and j, $X+j = i \cdot M'$ with $0 < i < 2^n/M'$ and $1 < j < M'$. He will recognize this by finding $KS(j) * \ldots * KS(j+b-1) = f(i \cdot M') * \ldots * f(i \cdot M'+b-1)$. (This will occur once by the assumption on b.) Thus he can conclude that $X = iM'-j$. This second stage takes $2^n/M'$ steps, so the entire algorithm takes $O(M+2^n/M')$ steps which is $O(2^n/M)$ for $M \leq 2^{n/2}$.

To break system B, the analyst proceeds as above, except that the second stage must be repeated for each possible value of K. This takes $O(2^{n+k}/M)$ steps total, assuming $M \leq \min(2^{(n+k)/2}, 2^n)$.

It is more efficient to use a somewhat smaller value of b (e.g., $b \sim (k+n)/c$ for a random f), allowing a small number of false alarms, and checking each one by composing more values.

Thus, for a random f, there is a tradeoff--more known plaintext (and memory) enables a proportional cut in exhaustive search. It is natural to wonder whether this attack is the best possible. However, lower bounds for the cryptanalysis of such a system with a specific f would have to include the size of the program, in order to eliminate the possibility of a table lookup. Furthermore, proving such results is far beyond the current theory of computational complexity. Nevertheless, it is easy to show that Theorem 1 is optimal if we restrict attention to attacks (as in Theorem 1) which are uniform in f.

We consider algorithms which have access to f only via an oracle, i.e., a black-box which outputs f(Y) whenever given Y. Since determining a value of f by calling the oracle requires at least one step, we can use simple information-theoretic arguments to get a lower bound on computation time.

Theorem 2

If such an oracle-algorithm for breaking system A has a significant probability (for f uniformly distributed over all functions of the appropriate type) of outputting the correct answer using M known blocks of keystream and calling the oracle W times, then $M \cdot W \sim 2^n$. For breaking system B, $M \cdot W \sim 2^{k+n}$ is required.

Proof

For system A, suppose the M values of the keystream $f(X+i_1)$, $f(X+i_2)$, ... $f(X+i_M)$ are known. Calling the oracle for a value $f(Y)$ may provide the answer immediately if $Y=X+i_j$, otherwise it merely eliminates the M possibilities: $X=Y-i_1$, $X=Y-i_2$, ... or $X=Y-i_M$, providing no information about any other possibility. Thus the probability that the oracle is ever asked for a value $f(X+i_j)$ is at most $W \cdot M/2^n$. If this does not occur, X must have a uniform distribution over the possibilities which have not been eliminated; there are at least $2^n - W \cdot M$ with these. Thus the probability of success is $\leq W \cdot M/2^n + 1/(2^n - W \cdot M)$. This is negligible for $W \cdot M << 2^n$.

Theorem 2 of course proves nothing about any specific function. However, it is reasonable to hope that a sufficiently complicated pseudo-random function, for instance a substitution-permutation network with enough rounds, will behave in a similar way. At least, what weaknesses do exist, should be hard to find, forcing the analyst to rely on uniform attacks of the sort covered by Theorem 2.

Theorem 2 suggests that our cryptanalytic attack is the best possible without using some special property of the basic function f. One such property is being a block code, i.e., $f(K,X)$ being an easily invertible permutation of f for each fixed K. Thus, while DES in counter mode is an example of system B, Theorem 2 does not apply. Hellman (1) shows how to attack that system using only two blocks of known keystream if c is large (i.e., if most of the ciphertext is used in the keystream). If c=64 (all the ciphertext is used) then his attack requires only $O(2^{56})$ operations [ours would require $O(2^{119})$]. For c=64-j (j bits are dropped from each block), he requires $O(2^{56+j})$. Thus his attack is better for large values of c, while ours is better for small values of c and large amounts of known keystream. It is an open question whether these two attacks are the best possible on a block code used in counter mode. "Best possible" would mean best for the black-box model of (2), in which the oracle can be asked for decryption as well as encryption.

Thus, the invertibility of a block code is a weakness when creating keystream ciphers. This suggests the desirability of adopting a standard one-way function. DES is a 120 bit to 56 bit function which is hard to invert given the last 64 bits of its argument, but easy to invert given the first 56. It should be possible to invent a genuine one-way function of comparable complexity, mapping 128 bits to 64 bits, which would be hard to invert given any subset of the input bits.

Such a function would provide a high degree of security when used in counter mode, as in system A. Using a secret 128 bit starting point X would provide 128-logM bit security, where M is the maximum number of 64 bit keystream blocks an opponent might conceivably obtain. Thus, if no more than a gigabyte was to be transmitted, using a single key, better than 100 bit security could be achieved.

Such a one-way function would also be useful to build a one-way hash function mapping data of arbitrary length to a fixed length. Such hash functions are used in digital signatures and password authentication. Finally, it might actually be easier to construct secure one-way functions via substitution-permutation networks without the restrictions created by the requirement of invertibility (e.g., the left-right division of DES, or the invertible S-boxes of Lucifer).

REFERENCES

(1) M. E. Hellman, "On DES Based Synchronous Encryption," to appear in *Computer*.
(2) M. E. Hellman, E. Karnin, and J. Reyneri, "On the Necessity of Cryptanalytic Exhaustive Search," Crypto 81.

USING DATA UNCERTAINTY TO INCREASE THE CRYPTO-COMPLEXITY OF SIMPLE PRIVATE KEY ENCIPHERING SCHEMES

G. M. Avis and S. E. Tavares

Department of Electrical Engineering

Queen's University

Computational efficiency is of prime importance to any microprocessor based cryptosystem. A technique is presented here which permits a reduction in the enciphering complexity of private key schemes without a loss in security. The net result can be a simplification of the system's implementation, a reduction in cryptographic overhead and the potential for a simple mathematical analysis of the security system.

Up to this time most modern enciphering schemes have been exceedingly complex in order to remain secure against chosen plaintext attack (CPA). It is apparent that there may be a more efficient method for defeating this formidable attack. Suppose that unpredictable (eg. random) bits are injected into the message stream before encryption. This unpredictable data is generated internally by the cryptosystem and is kept secret from the external world. This denies the cryptanalyst a CPA on the enciphering scheme. Indeed, a CPA on the cryptosystem may allow the choice of some of the bits submitted to the enciphering function, but the injected bits will remain completely unknown. The addition of this 'data uncertainty' essentially nullifies the power of the CPA and greatly restricts the cryptanalyst's access to the enciphering algorithm.

In order to demonstrate the concept of data uncertainty, consider a very simple (and highly insecure) enciphering function:

$$C = X^E \bmod N$$

where: C – ciphertext
X – plaintext block (X < N)
E, D, N integers such that:

$$X = C \cdot D \bmod N$$

$\text{GCD}\,[E, N] = 1$
$E \cdot D = 1 \bmod N.$

The secret key consists of the integers (E, D, N). Very little analysis is required to discover that this procedure is completely unacceptable as a standalone enciphering function. However, the story is different when data uncertainty is added in the following manner:

Let $X = R \,||\, M$ where: X – as before
R – random bit string (secret)
M – message block
$||$ – concatenation operation

Data uncertainty is introduced by appending a random bit string onto the front of the message block. The random string is generated secretly by the cryptosystem and is never revealed to the external world. Decryption at the receiver is not hampered in any way, the random bit string is simply discarded before the message is output by the decryption box. The general form of the system is given in fig. 1. For this example, the function 'U' would simply be concatenation and 'f_K' would be the enciphering function given above.

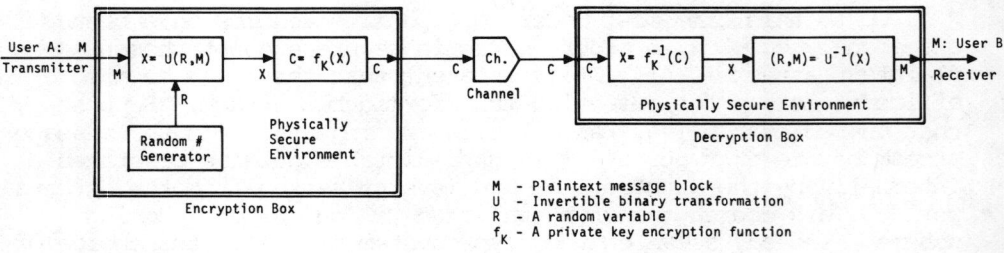

M – Plaintext message block
U – Invertible binary transformation
R – A random variable
f_K – A private key encryption function

Figure 1

Data uncertainty

Some interesting mathematical analysis can be performed upon this cryptosystem. Let $f(X, E, N)$ be the enciphering function applied to 'X'. Hence:

$$
\begin{aligned}
f(X,E,N) &= X \cdot E \bmod N \\
&= (R \| M) E \bmod N \\
&= (R \cdot 2^v) E \bmod N + X \cdot E \bmod N \\
&= f(R,E',N) + f(X,E,N) \quad (1)
\end{aligned}
$$

In (1) it is clear that the encryption can be broken down into the sum of two other encryptions: one is simply the encryption of the message without data uncertainty; the other is a function of a random variable. Mathematically this is a very interesting construct, somewhat similar to the one time pad (though not equivalent of course). Despite its simplicity the system seems to be secure for reasonably large key sizes. Further investigation into this hypothesis would be welcomed.

The major drawback of data uncertainty is that it causes bandwidth expansion in the message stream. To counter this problem is is possible to combine the data uncertainty idea with a security protocol. Typically a secure message exchange system would employ security information like sequence numbers, time stamping or CRC checks in order to allow the receiver to perform some data verification [4]. This information can be used as data uncertainty provided that it is 'randomized' before being appended to the message block (figure 2).

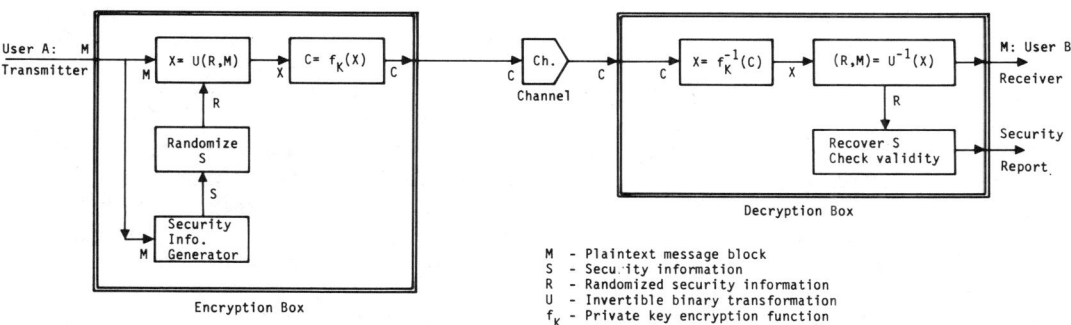

Figure 2

Combining data uncertainty and security information

The randomization process must be invertible at the receiver and unpredictable by the external world. The easiest realization would be to encrypt the security information before appending it to the message block.

It is apparent that the data uncertainty scheme proposed here could be susceptible to a chosen ciphertext attack.[1] It is possible for an active wirewrapper to inject ciphertext onto the channel which would cause the receiver to reveal its secret key. In order to prevent this, the receiver must be able to authenticate incoming ciphertext so that only legitimate messages emerge to the external world. Any reasonable secure message exchange protocol would ensure that this authentication takes place. However, it is interesting to note that such authentication is indeed necessary for the security of the system.

A prototype of this system has been implemented in software alongside the RSA cryptosystem [7]. The code executes on a Texas Instruments 9900 microprocessor, a 16-bit machine with a 3-MHz clock rate. A number of practical advantages have been noted. The private key system can maintain a higher throughput than a software version of the Data Encryption Standard (DES) running on the same machine (in fact, it is faster than a number of special purpose chips implementing DES). Another advantage is that the private key system can share the majority of its code with the RSA algorithm. Once RSA has been implemented, the addition of the private key system requires very little extra code. This sharing could be critical if a "hybrid" cryptosystem were implemented on a single VLSI chip. The desirability of combining public and private key cryptography has been well documented [1,2,5,6]. Implementing a single chip hybrid cryptosystem would be a major cryptographic milestone.

It should be clear that considerable benefit can be derived from the use of data uncertainty. The idea of injecting random noise into data prior to encryption has already been recognized in the literature [3], however, it is possible that the concept has not yet been exploited to the fullest degree. The example illustrates that even very simple enciphering functions can be strengthened using this technique. Hence cryptosystems can be created which are mathematically simple to analyze: in the example above cryptanalysis can be reduced to a problem of simultaneous equations. This could lead to a derivable measure of the system's cryptocomplexity. Data uncertainty can be realized in any number of ways and as a concept should provide good potential for further study [1]. The example presented here is intended mainly to serve as a demonstration that brute force enciphering functions may not be the most efficient method for concealing data. A "layered" approach to enciphering may well be the best way to go [8].

1. Preliminary comments by Ronald Rivest (M.I.T.) and Andrew Yao (U.C. Berkeley).

BIBLIOGRAPHY

1. Avis, G., "A MICROPROCESSOR BASED SECURE MESSAGE EXCHANGE SYSTEM", M.Sc. Thesis, Dept. of Electrical Engineering, Queen's University, Kingston, Ontario, Canada, June, 1982.
2. Akl, S., "ON DIGITAL SIGNATURES WITH BLINDFOLDED ARBITRATORS WHO CANNOT FORM ALLIANCES:, Proceedings of the 1982 Symposium on Security and Privacy, IEEE, Oakland, CA, Apr.'82
3. Davida, G., Well, D., Kam, J., "A DATABASE ENCRYPTION SCHEME WITH SUBKEYS", ACM Trans. on Database Systems, Vol. 6, No. 2, June 1981.
4. DeMillo, R., "CRYPTOGRAPHIC PROTOCOLS", American Math Society, Cryptography in Revolution: Mathematics and Models, Jan. 1981.
5. Kowalchuk, J., Schanning, B. & Powers, S., "COMMUNICATIONS PRIVACY: INTEGRATION OF PUBLIC AND SECRET KEY CRYPTOGRAPHY", MITRE Corp., Bedford, MA, June 1980.
6. Lau, Y. & McPherson, T., "THE IMPLEMENTATION OF A HYBRID RSA/DES KEY MANAGEMENT SYSTEM", CRYPTO '81, UCSB, CA, Aug. 1981.
7. Rivest, R., Shamir, A. & Adleman, L., "A METHOD FOR OBTAINING DIGITAL SIGNATURES AND PUBLIC KEY CRYPTOSYSTEMS", Comm. of the ACM, Vol. 21, No.2.
8. Spencer, M. & Tavares, S. "LAYERED ENCRYPTION APPLIED TO BROADCAST SYSTEMS", Dept. of E.E., Queen's University, Kingston, Ontario, Canada, Feb. 1982.

RANDOMIZED ENCRYPTION TECHNIQUES[1]

Ronald L. Rivest and Alan T. Sherman
MIT Laboratory for Computer Science
Cambridge, MA 02139

Abstract

A *randomized encryption procedure* enciphers a message by randomly choosing a ciphertext from a set of ciphertexts corresponding to the message under the current encryption key. At the cost of increasing the required bandwidth, such procedures may achieve greater cryptographic security than their deterministic counterparts by increasing the apparent size of the message space, eliminating the threat of chosen plaintext attacks, and improving the *a priori* statistics for the inputs to the encryption algorithms. In this paper we explore various ways of using randomization in encryption.

I. Introduction

Cryptographers often enhance the security of their codes and ciphers by using randomization in the encryption process.[2] For instance, it is common practice to define codes by associating with each plaintext word a set of codewords; to encode a plaintext word, the encoding procedure randomly selects a codeword from the corresponding set. An elegant example of this idea is the Jefferson-Bazeries Wheel Cipher and its variations, ciphers used by the U.S. during both world wars [Kru81], [Kah67]. Another common randomized encryption technique is to intersperse null

---

[1] This research was supported by NSF grant MCS-8006938.
[2] We assume the reader is familiar with the general principles of cryptology — as presented in [Dil79], for example.

plaintext words in the input stream or to intersperse null codewords in the output stream of an encryption algorithm.

New randomized encryption techniques have been recently published by Wyner [Wyn75], McEliece [McE78], Lempel [Lem79], Sloane [Slo82], Nicolai [Nic82], and Goldwasser and Micali [GoM81,82]. Randomization is also proposed to achieve secure digital signatures in Rabin's variation of the RSA cipher [Rab79], to enlarge the apparent size of the message space in the encryption of passwords under the *Unix* operating system, and to create the *join* of two cryptosystems — an encryption scheme as strong as the stronger of two component encryption schemes [AsB82]. This paper is strongly motivated by these examples.

The goal of randomized encryption is increased security, and this goal may be achieved through each of the following means:
- Smoothing out the *a priori* statistics for the distribution of inputs to the encryption algorithm or one-way function. This may eliminate or reduce the effectiveness of statistical attacks.
- Eliminating the possibility of chosen plaintext attacks. If the encryption function encrypts only randomly generated bit sequences, a chosen plaintext attack is impossible to mount.
- Increasing the apparent size of the message space to the enemy. If the message space is small, a non-randomized scheme runs the risk of being defeated by simple statistical or forward search [Sim82] attacks.

In information theoretic terms, randomized encryption attempts to obtain higher levels of security through increasing the entropy of the plaintext [Sha49], [Gal68]. In this respect, randomized encryption is similar to source coding. However, while source coding reduces redundancy through message compression, randomized encryption increases entropy by adding random bits. Although randomized encryption increases the bandwidth, it is a relatively simple technique that can be easily applied in many applications.

In this paper we explore how randomization can be used during encryption. In section II, we explain what randomized encryption is. Next, in section III, we categorize many ways randomization can be used. In section IV, we discuss extensions and applications of randomized encryption, and, in section V, we discuss directions for further research. We summarize our results in section VI.

II. Randomized encryption

A *randomized encryption procedure* produces ciphertext via a nondeterministic function of the message to be encrypted and the encryption key. For each key, a given message may be encrypted in several ways; the encryption procedure makes a random choice to decide which way to use. We assume, though, that encryption is uniquely decodable: for each key, each ciphertext corresponds to at most one message.

More precisely, a randomized encryption procedure is a relation $\Pi \subseteq M \times K \times C$, where M is the message space, K is the key space, and C is the ciphertext space, such that for each key $k \in K$ and each ciphertext $c \in C$, there is at most one $x \in M$ such that $(x, k, c) \in \Pi$. Also, for each message $x \in M$ and each key $k \in K$ there is at least one ciphertext $c \in C$ such that $(x, k, c) \in \Pi$. The quadruple (M, K, C, Π) is called the *randomized encryption system*.

In practice, a randomized encryption procedure would encrypt a message x under a key k as follows. First, a bit sequence r would be randomly chosen from a set R of bit sequences. Second, the value $\psi(r, x, k)$ would be computed, where ψ is a deterministic function such that $\psi : R \times M \times K \mapsto C$ and $\Pi = \bigcup_{r \in R, k \in K, x \in M} (x, k, \psi(r, x, k))$.

Randomized encryption schemes require a source of random bits. Such a source may be built using neon discharge tubes, noisy diodes [Mad72], radioactive decay [Kle60], or other natural sources of randomness [Gif82]. We assume such generators are capable of producing unbiased and totally unpredictable bit sequences as rapidly as needed. Pseudo-random bit sequence generators, such as the one proposed by Blum and Micali [BlM82], are not suitable replacements for random bit sources here.

Since, for any key, each message corresponds to several ciphertexts and each ciphertext corresponds to at most one message, the ciphertext space will be larger than the message space. This requires a certain amount of *bandwidth expansion* in the communication channel since more bits must be transmitted to specify the ciphertext than are needed to identify the message. Bandwidth expansion is the major cost of using randomized encryption, and it is unavoidable. As a measure of this cost, we define the *bandwidth expansion factor* of an encryption procedure to be the ratio of the the number of ciphertext bits transmitted to the corresponding number of message bits. If the bandwidth expansion factor is not constant between keys or even between messages, a more general notion of an *average bandwidth expansion factor* could be used instead. Most of the schemes reviewed in this paper have uniformly constant bandwidth expansion.

Figure 1 shows the block diagram for the secure transmission of data over an insecure communications line using a randomized encryption procedure. Note that the physical random bit sequence generator is within the secure area of the encryption unit. We assume an enemy cannot determine the intermediate results of the encryption and decryption computations. In particular, the output of the

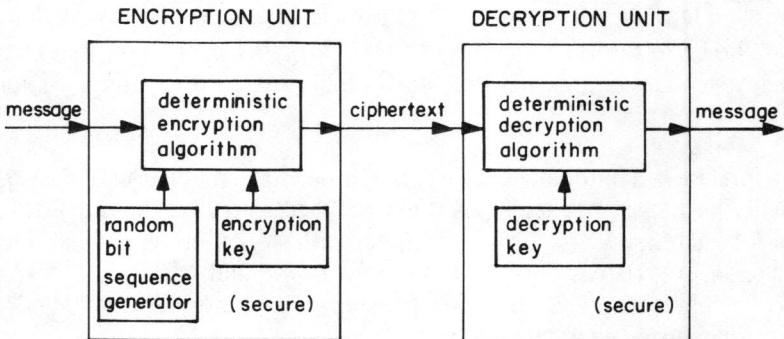

Figure 1. Block diagram of a randomized encryption procedure

physical random bit sequence generator is hidden from the enemy, unless it is part of the ciphertext.

III. Classification of randomized encryption methods

In this section, we will present several ways in which randomization can be used in encryption. We organize the schemes in two categories: randomized block ciphers and randomized stream ciphers.[3]

III.A. Notation

We use the following notation:

$\mathcal{M}, \mathcal{C}$ denote the message space and ciphertext space of a randomized encryption algorithm.

$\mathcal{K}$ denotes a set of keys.

$\mathcal{R}$ denotes a set of bit sequences.

M denotes a message source that generates messages in $\mathcal{M}$.

---

[3] A *stream* or *finite state* cipher is based on an encryption function that can be in one of finitely many different states. First, the state is set to some initial value; then, after each application of the cipher, the state is updated by some *state transition function*. The encryption process depends on the message, the key, and the current state. A *block* or *memoryless* cipher is a cipher that does not depend on any state information.

Randomized Encryption Techniques

R, R' denote sources of random bits that generate sequences in $\mathcal{R}$.

k, k' denote cryptographic keys in $\mathcal{K}$.

$|M|$ and $|R|$ denote the lengths of messages and bit sequences generated by M and R respectively.

E, E' denote deterministic encryption functions mapping $\mathcal{K} \times Domain(E)$ into $Target(E)$, where $Domain(E)$ is the message space of E and $Target(E)$ is the ciphertext space of E. For many examples it will be true that $M = Domain(E) = Target(E) = \{0,1\}^n$, for some n. When E is subscripted, the subscript refers to the key in use: thus, for each key $k \in \mathcal{K}$, E_k denotes an injective map from $Domain(E)$ into $Target(E)$.

D, D' denote decryption functions corresponding to E, E'. For any message x, and any key k, it must be true that $D_k(E_k(x)) = x$.

F, F' denote key-dependent deterministic one-way functions mapping $\mathcal{K} \times Domain(F)$ into $Target(F)$. Note that F_k need not be injective for any key k.

P, P' denote pseudo-random bit stream generators.

We will use the descriptive but slightly abusive notation $E_k(R)$ to denote an encryption scheme in which the sequences generated by the random source R are encrypted with encryption function E under key k. When the meaning is clear, we will also abbreviate $E_k(R)$ by $E_k R$ or ER. We will also use similar notations involving M, D, and F.

$\|$ denotes concatenation. For example, $a \| b \| c$ denotes the concatenation of the strings a, b, c.

$\oplus$ denotes exclusive or (XOR).

For example, $EM \| R$ denotes the scheme in which the final ciphertext is obtained by concatenating (1) the ciphertext obtained by enciphering M with encryption function E, and (2) a randomly generated block of bits.[4] Although this scheme increases the required bandwidth for no apparent benefit, it can serve as a useful reference when considering other schemes.

When evaluating a randomized encryption scheme it is useful to distinguish between intrinsic weaknesses of the scheme and weaknesses due to bad component

---

[4] We adopt the convention that function application takes precedence over concatenation. Thus, $(EM \| R) = ((EM) \| R)$.

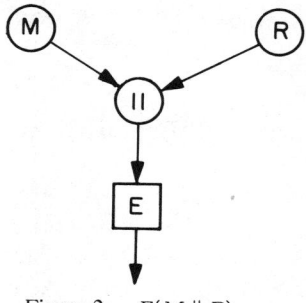

Figure 2. $E(M \parallel R)$

cryptographic functions. An *intrinsic* weakness occurs regardless of what component functions are used.

III.B. Randomized block ciphers

In this section, we review a number of ways of using randomization in block ciphers. We assume that the sizes of the message and ciphertext blocks are fixed parts of each encryption system.

III.B.1. $E(M \parallel R)$

Perhaps the simplest randomized encryption technique is to concatenate a random sequence[5] to the message and then encrypt the result. The decryption unit applies the decryption function and then discards the random sequence. We call this method the $E(M \parallel R)$, or *random padding* technique. See Figure 2.

There are five advantages of the $E(M \parallel R)$ scheme. First, this scheme increases the apparent size of the message space to the enemy in the sense that $Domain(E)$ is $M \times R$ rather than M. Second, under a chosen plaintext or chosen ciphertext attack, this scheme does not allow the enemy to obtain any complete input to the encryption function. Therefore this scheme partially protects against such attacks. Third, this scheme somewhat improves the distribution of inputs to the encryption function in that part of each input to the encryption function is randomly selected. Fourth, this scheme is not intrinsically vulnerable to *bit twiddling* — an active attack in which the enemy selectively modifies ciphertext bits in order to alter corresponding plaintext bits. Finally, since R may be chosen to be any size, this scheme allows for variable bandwidth expansion. Of course, if random padding is to do much good, R must be sufficiently large. If $|M| = |R|$, then the bandwidth expansion factor is two.

---

[5] For brevity, we use the phrase *random sequence* to mean randomly generated bit sequence.

At least one provably secure public-key cryptosystem fits this classification. In particular, the Goldwasser/Micali scheme [GoM81,82] can be viewed as an $E(R \parallel M)$ scheme, where $E(x) = g^x \pmod{n}$; n is a large composite number; $|M| = 1$; and g is a generator of Z_n^*, the multiplicative group modulo n. In the Goldwasser/Micali scheme, each message is sent one bit at a time. The security of their scheme is based on the assumption that it is difficult to determine in general whether or not an integer z is a quadratic residue modulo n, where $0 < z < n$ [NiZ80]. To encrypt a 0, the sender randomly selects a quadratic residue modulo n. Similarly, to encrypt a 1, the sender randomly selects a quadratic non-residue modulo n with Jacobi symbol equal to 1. Here n is a large composite number published by the intended recipient of the message. To enable the sender to encrypt, the intended receiver also publishes a quadratic nonresidue y modulo n with Jacobi symbol equal to 1. Since the intended recipient knows the factorization of n, he can determine whether the transmitted numbers are quadratic residues modulo n. Randomization is essential to the security of the system: unless the encryption of each bit is randomly chosen, the enemy could break the system. Unfortunately, the bandwidth expansion factor of the Goldwasser/Micali scheme is several hundred, making it impractical for many applications.

Avis and Tavares [AvT82] present a similar example of a simple cipher that they conjecture is hard to break when used in an $E(R \parallel M)$ mode. For the Avis and Tavares scheme, $E(x) = xe \pmod{n}$, where n is a large composite number and e is relatively prime to n. The security of this scheme remains to be tested.

III.B.2. $(M \oplus R) \parallel ER$

The $(M \oplus R) \parallel ER$ scheme is similar to the Vernam one-time pad in that the message is XORed with a randomly generated bit sequence. But unlike the one-time pad, the random sequence is encrypted and then concatenated to the enciphered message. Asmuth and Blakley refer to this scheme as the cryptographic exponential of the underlying cryptosystem [AsB82]. See Figure 3.

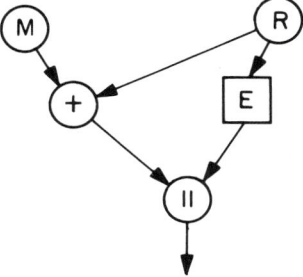

Figure 3. $(M \oplus R) \parallel ER$

The main idea of this scheme is to apply the encryption function only to random bit sequences, guaranteeing uniform *a priori* statistics of the inputs to the encryption function. Whereas the previous method uses the encryption function to mix the random bits with the message bits, this scheme directly randomizes each bit of the message. There are, however, some possible weaknesses with this technique. First, as with the one-time pad, there is the danger with this scheme that an active wiretapper could selectively change certain bits of the plaintext by modifying the corresponding bits of the ciphertext. Second, a chosen ciphertext attack against the system amounts to a chosen ciphertext attack against the encryption function. (See section IV.A. for ways to prevent such attacks.) Third, although the inputs to E are randomly selected, the conditional probability distribution $Prob(R \mid M \oplus R)$ is identical to the *a priori* probability distribution of the message space. Finally, the scheme is vulnerable to a spoofing attack in which an active wiretapper interjects spurious messages into the communications channel: to forge messages, the enemy uses a known plaintext attack to obtain a pair of numbers r, y such that $E(r) = y$. Assuming $|M| = |R|$, this method also has a bandwidth factor of two.

III.B.3 $E_R(M) \parallel E'R$

The $E_R(M) \parallel E'R$ technique encrypts the message with a random message key that is encrypted under the current session key. Note that the $(M \oplus R) \parallel ER$ scheme is also special case of the $E_R(M) \parallel E'R$ technique, where E is the Vernam one-time pad. See Figure 4.

Provided $|M|$ is much larger than $|R|$, the bandwidth expansion factor can be quite small. The main advantage of this scheme is to reduce the threat of attacks on the the encryption function E that require many ciphertexts produced under the same key.

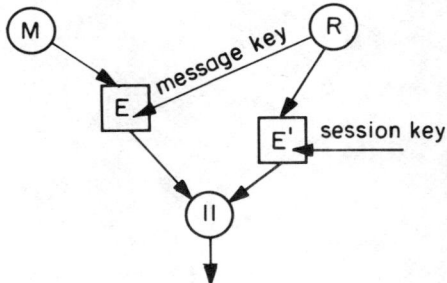

Figure 4. $E_R(M) \parallel E'R$

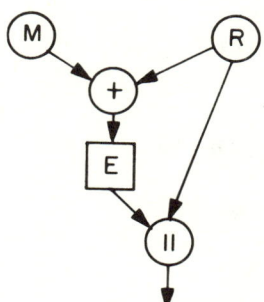

Figure 5. $E(M \oplus R) \| R$

III.B.4. $E(M \oplus R) \| R$

The $E(M \oplus R) \| R$ technique XORs each message with a random bit sequence and then encrypts the result. The random bit sequence is sent in the clear. See Figure 5.

This method has bandwidth expansion factor two and has all of the cryptographic properties mentioned for the $(M \oplus R) \| ER$ scheme.

III.B.5. $E((M \oplus R) \| R)$

The $E((M \oplus R) \| R)$ method combines some of the ideas from the random padding scheme and the $(M \oplus R) \| ER$ scheme. This method XORs each message with a random bit sequence and encrypts the result concatenated with the random sequence. See Figure 6.

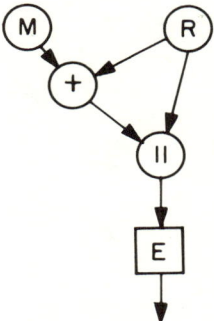

Figure 6. $E((M \oplus R) \| R)$

This scheme increases the apparent size of the message space to the enemy, smoothes out the distribution of inputs to the encryption function, and is not intrinsically vulnerable to bit twiddling. Furthermore, this scheme intrinsically allows neither a chosen plaintext attack nor a chosen ciphertext attack against the encryption function. One potential weakness of this scheme is that under a chosen plaintext attack the enemy can force the encryption function to be applied to bit sequences of the form $r \parallel r$.

III.B.6. $(M \oplus FR) \parallel R$

Finally, we consider the $(M \oplus FR) \parallel R$ method. This method is similar to the $(M \oplus R) \parallel ER$ technique, except that the random sequence is transmitted in the clear and XORed into each message in encrypted form. Unlike the $(M \oplus R) \parallel ER$ technique, this method does not require the cryptographic function to be invertible. Asmuth and Blakley call this scheme the cryptographic exponential of a key dependent one-way function [AsB82]. See Figure 7.

For this scheme, a chosen plaintext attack amounts to a known plaintext attack against the cryptographic function F and a chosen ciphertext attack amounts to a chosen plaintext attack against F. Note that the way in which the $(M \oplus FR) \parallel R$ method applies F and combines M, FR, and R is similar to the operations performed in each round of DES [FIP77]. DES, however, does not involve randomized encryption.

We will now show that to every $(M \oplus R) \parallel ER$ scheme there is an equivalent $(M \oplus FR) \parallel R$ scheme, in the following sense. We say that two randomized encryption systems are *equivalent* if (1) they have the same message space (including probability distribution), keyspace, and ciphertext space, and (2) for each key and message, the two randomized encryption procedures produce the same set of ciphertexts with the same probabilities. Although equivalent randomized encryption systems may not produce the same ciphertext for a given randomly generated bit

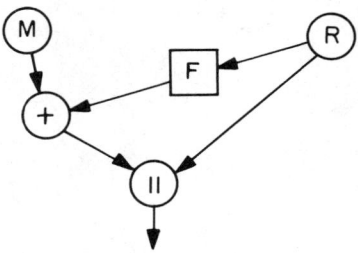

Figure 7. $(M \oplus FR) \parallel R$

sequence, they will yield the same set of ciphertexts with exactly the same probability distribution. Thus, equivalent randomized encryption systems are cryptographically identical in that the enemy cannot distinguish between them.

Proposition 1

Let $\mathcal{E}$ be any $(M \oplus R) \parallel ER$ randomized encryption system such that $M = R$ and such that, for each key k, the encryption function E_k permutes M. Then, there is a $(M \oplus FR) \parallel R$ randomized encryption system equivalent to $\mathcal{E}$.

Proof.

Since, for each key k, E_k permutes M, it follows that $(M \oplus R) \parallel E_k R$ and $(M \oplus D_k R) \parallel R$ are equivalent randomized encryption systems. Hence, by choosing $F = D$, we are done.   Q.E.D.

III.B.7. Schemes based on block error-correcting codes

In this section, we consider randomized block ciphers based on error-correcting codes. Here, G denotes an error-correcting code that is capable of correcting at most t errors, and ϵ denotes a source of random bit sequences with at most t ones. The bandwidth expansion factors of these schemes depend on what particular error-correcting codes are used.

III.B.7.a. $GM \oplus \epsilon$

In the $GM \oplus \epsilon$ scheme, the message is first encoded by a secret error-correcting code. Then, a randomly selected subset of the bits in the encoded message is changed. The number of bits changed is at most the error-correcting capacity of the code.

This scheme is based on a public-key cryptosystem proposed by McEliece [McE78]. In McEliece's system, the sender first encodes a message by using a public generator matrix for an error-correcting code. The sender then changes a randomly selected subset of the the bits of the codeword. Provided the sender changes no more bits than the error-correcting code is capable of correcting, the receiver will be be able to decode the message. To decode, the receiver uses a secret trapdoor error-correcting code related to the public generator matrix. The security of the system is based in part on the NP-completeness of the general decoding problem for linear error-correcting codes [BMT78]. The security also depends on the sender randomly selecting what bits to change.

III.B.7.b. $E(GM \oplus \epsilon)$

As with random padding, in the $E(GM \oplus \epsilon)$ method, random bits are combined with the message and then the resulting block is encrypted. In this scheme random bits are combined with the message by first encoding the message with an error-correcting code and then altering a randomly selected subset of bits in the resulting codeword. The difference between random padding and the $E(GM \oplus \epsilon)$ scheme is in the way in which the random bits are combined with the message. In this method, G need not be secret.

III.B.8. Wire-tap channel schemes

Randomized encryption appears to be especially powerful for situations in which the enemy must listen to a noisy communications channel known as the *wire-tap channel*. As an example, we will review a simple scheme described by Sloane [Slo82]. This example illustrates some of the important ideas in Wyner's research on the wire-tap channel [Wyn75].

Sloane [Slo82] describes a secure way to send one bit over an insecure channel under the assumption that the enemy is forced to listen to a noisy line. To encrypt a bit, the sender randomly selects a bit sequence whose parity is equal to the message bit. The sender chooses the bit sequence long enough so that, due to the noise in the wiretap channel, the enemy will be unable to determine the parity of the codeword.

III.C. Randomized stream ciphers

Since the message of a stream cipher may be viewed as a stream of characters or bits to be encrypted one at a time, stream ciphers provide a convenient context in which to describe encryption techniques that intersperse nulls into the plaintext or ciphertext streams. Furthermore, because the encryption process of a stream cipher depends on the current state of the cipher, randomization can be applied not only to the message, but also to the state. We will describe three randomized techniques for stream ciphers: first, we will discuss two methods for interspersing nulls into the input and output streams; and, finally, we will propose a technique for constructing a randomized stream cipher that uses a nondeterministic state transition function. For another interesting randomized stream cipher based on a nondeterministic state transition function, see Carl Nicolai's paper entitled "Nondeterministic Cryptography" [Nic82].

Many stream ciphers use a pseudo-random stream generator. Such a pseudo-random stream generator may be a nonlinear feedback shift register, for example. All of the randomized techniques described in this section will be discussed in terms of pseudo-random stream generators whose keys are known to both sender and receiver. A class of ciphers related to the randomized stream ciphers presented here can be obtained by replacing the pseudo-random bit generators with cipher-feedback arrangements.

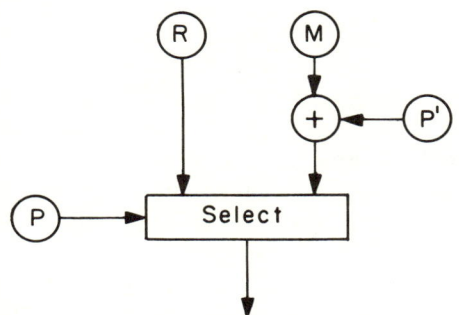

Figure 8. (if $P = 0$ then R else $(M \oplus P')$)

III.C.1. (if $P = 0$ then R else $(M \oplus P')$)

This scheme uses two pseudo-random bit stream generators P, P' and provides a way to intersperse nulls into the output stream. One of the pseudo-random bit stream generators is used to encrypt the message, while the other is used to select where random bits are to be interspersed into the ciphertext stream. Specifically, to generate the next ciphertext bit, a pseudo-random bit is taken from the generator P. If this bit is 0, the next ciphertext bit is obtained from the random bit generator R. Otherwise, the ciphertext bit is the XOR of the next message bit and the next output bit from the pseudo-random bit stream generator P'. We call this method *pseudo-randomly interspersing random bits after encryption*. See Figure 8.

III.C.2. $(R \oplus P) \parallel ((\text{if } R = 0 \text{ then } R' \text{ else } M) \oplus P')$

In this scheme, a random bit generator R is used to select where random bits are to be inserted into the message stream. The resulting stream is then encrypted by a pseudo-random bit stream generator P'. To enable the receiver to decipher the message, the bit sequence generated by R is also sent, encrypted by another pseudo-random bit stream generator P. We call this method *randomly interspersing random bits before encryption*. See Figure 9. Although this scheme and the previous one look promising, they have not yet undergone critical analysis.

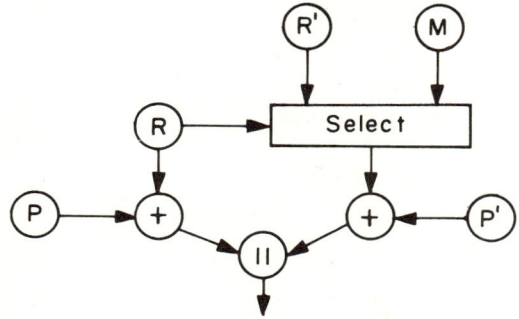

Figure 9. $(R \oplus P) \parallel ((\text{if } R = 0 \text{ then } R' \text{ else } M) \oplus P')$

III.C.3. Random walk

We now propose a randomized stream cipher that is based on a a nondeterministic state transition function. In this scheme the encryption unit maintains a register of length n holding its current state. On each cycle, the state is updated and one message bit is encrypted and transmitted. To update the state, the state register is left-shifted one position, and a new truly random bit is brought in as the new rightmost bit of the state register. Each message bit is encrypted by XORing it with the value of a binary-valued function F applied to the current state. We call this method the *random walk* scheme because the encryptor takes a random walk in the state space.

To help the intended receiver decipher the message, the encryptor also sends with each encrypted message bit a hint about the current state. This hint is the result of applying a ternary-valued function F' to the current state. Thus, for each message bit, the corresponding ciphertext consists of one binary digit and one ternary digit. Decryption is still tricky since the intended receiver cannot always be certain what state transition took place in the encryption unit. One idea behind this scheme is the conjecture that cryptographic security can often be enhanced by making it difficult for even the intended receiver to decipher the message.

Assuming that F' yields each of the three possible output values equally often, it is possible to argue that the receiver will be able to adequately follow the encryptor's progress through state space. In some cases, however, the decoder may have to wait several cycles before being able to decode a given bit. Fortunately, since the transmitter can keep track of the degree of uncertainty in the receiver, the transmitter can avoid buffer overflow problems in the receiver by transmitting hints without ciphertext bits if necessary whenever the receiver happens to arrive at a situation of high ambiguity. It can be shown that binary hints are unsatisfactory for

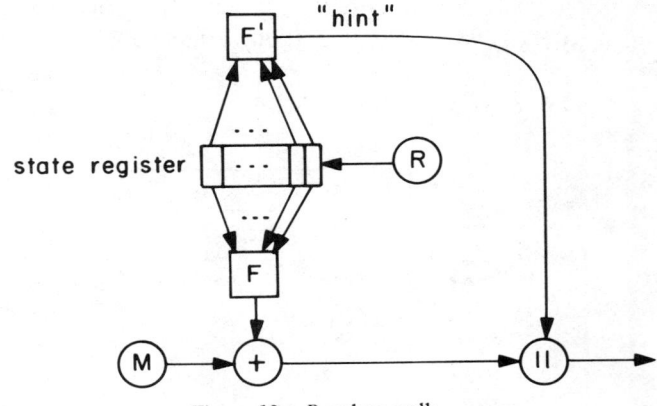

Figure 10. Random walk

this scheme. See Figure 10. Again, we urge the reader to attempt to cryptanalyze this scheme.

IV. Extensions and applications

In this section, we will describe a method for presenting chosen ciphertext attacks and discuss how randomization can be used to combine cryptosystems.

IV.A. Preventing chosen ciphertext attacks

Although most of the schemes presented in section III.B. prevent the enemy from mounting chosen plaintext attacks against the underlying cryptographic functions, many of the schemes do not prevent chosen ciphertext attacks. For example, a chosen plaintext attack against the $(M \oplus R) \parallel ER$ method amounts only to a known plaintext attack against the encryption function E, but a chosen ciphertext attack against the method amounts to a chosen ciphertext attack against E. Similarly, a chosen plaintext attack against the $(M \oplus FR) \parallel R$ scheme amounts to a known plaintext attack against F, but a chosen ciphertext attack against the scheme amounts to a chosen plaintext against F. These chosen ciphertext attacks against the encryption schemes can be prevented by the standard technique of constructing the decryption unit so that it will output only properly constructed messages. For example, the decryption unit could be built so that it would output only messages that have valid cryptographic checksums appended to them, as illustrated in Figure 11. Because the enemy does not know the key to F, he should be unable to construct ciphertexts that would decrypt to messages with valid checksums, and should be unable to get the decryption unit to produce outputs. While chosen plaintext attacks can be prevented by adding random bits to the message, chosen ciphertext attacks can be avoided by adding redundancy to the message.

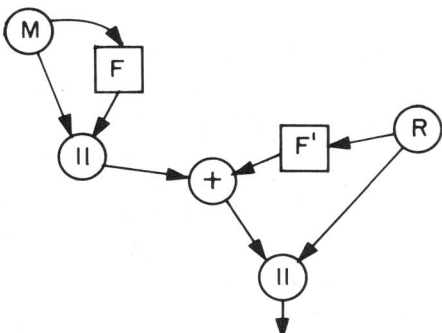

Figure 11. $((M \parallel FM) \oplus F'R) \parallel R$

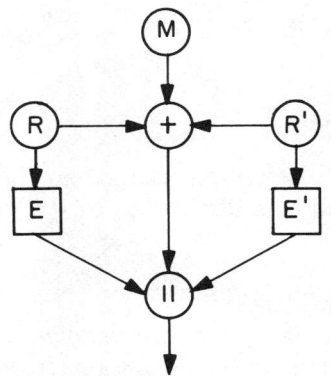

Figure 12. $(M \oplus R \oplus R') \parallel ER \parallel E'R'$

IV.B. Combining Cryptosystems

Randomized encryption techniques provide powerful tools for combining cryptosystems. For example, Asmuth and Blakley's [AsB82] ingenious method for combining two cryptosystems E and E' depends crucially on randomized encryption. In our notation, their system is $(M \oplus R \oplus R') \parallel ER \parallel E'R'$, which may be viewed as a generalization of the $(M \oplus R) \parallel ER$ method. See Figure 12. Asmuth and Blakley conjecture that, provided the keys for the two component cryptosystems E and E' are independently selected, the resulting randomized cryptosystem is at least as hard to break as either E or E'. Unfortunately, Asmuth and Blakley were unable to rigorously prove their conjecture. We consider it an interesting open question to provide a more formal statement and proof of the Asmuth/Blakley conjecture.

V. Directions for further research

We suggest several avenues for further investigation. First, the effect of randomized encryption techniques on specific cryptographic functions merits further study. Second, schemes involving multiple encryption functions and multiple random sources could be more thoroughly investigated. Third, it would be interesting to determine what operations other than XOR and $\parallel$ would be good methods for combining messages with random sequences. Among other operations, threshold schemes [Bla80] might be useful tools. Finally, it would be interesting to explore two-party randomized encryption systems. Known examples include Merkle's puzzle system [Mer78], Rabin's signature scheme [Rab78], Shamir, Rivest, and Adleman's mental poker protocols [SRA79], Rabin's oblivious transfer [Rab81], and Blum's protocol for exchanging secrets [Blu82].

VI. Summary

In this paper we have described many patterns in which randomization can be used in encryption for the purpose of enhancing cryptographic security. In the process, we have attempted to provide a unified framework against which recently discovered randomized cryptographic systems can be viewed. Although the schemes presented here represent only a small fraction of the available techniques, many of the methods not mentioned in this paper can be constructed by combining and extending ideas discussed here. We hope that this paper will stimulate the reader to think about additional ways in which randomization can be used.

While this paper has attempted to provide a rough taxonomy of the available techniques, the use of randomization itself deserves much further careful study. In particular, it would be interesting to determine for various encryption functions how much cryptographic security is gained by different randomization techniques. This is a difficult question, however, because, to answer it, we must first thoroughly understand what cryptographic security means. Of course, whether or not the enhanced security of randomized encryption is worth its associated cost of bandwidth expansion will depend on the particular application. Fortunately, advances in communications technologies (*e.g.* fiber-optics) may help alleviate the drawbacks of increased bandwidth. We believe that randomization has been and will continue to be an important tool in cryptographic design.

VII. Acknowledgments

We would like to thank David Gifford for reading the manuscript and making several useful comments. In addition, we are grateful to Joseph Shipman and several of the CRYPTO 82 attendees who pointed out some examples of randomized encryption systems.

VIII. References

[AsB82] Asmuth, C. A., and G. R. Blakley. An efficient algorithm for constructing a cryptosystem which is harder to break than two other cryptosystems. *Comp. & Maths. with Appls.*, 7 (1981), 447–450.

[AvT82] Avis, G. M., and S. E. Tavares. A microprocessor based cryptosystem for secure message exchange. *Advances in Cryptology: Proceedings of CRYPTO 82*, Plenum Press, (New York, 1983).

[BMT78] Berlekamp, E. R., R. J. McEliece, and H. van Tilborg. On the inherent intractability of certain coding problems. *IEEE Trans. on Info. Theory*, **IT-24** (1978), 384–386.

[Bla80] Blakley, G. R. The Vernam one-time pad is a key safeguarding scheme, not a cryptosystem. *Proceedings of the 1980 IEEE Symposium on Security and Privacy*, (1980), 447–450.

[BlM82] Blum, Manuel, and Silvio Micali. How to generate cryptographically strong sequences of pseudo random bits. *Proceedings of the 23rd Annual Symposium on Foundations of Computer Science*, (November, 1982), 112–117.

[Blu82] Blum, Manuel. How to exchange (secret) keys. *Proceedings of the 15th Annual ACM Symposium on Theory of Computing*, (May 1983), to appear.

[DiH79] Diffie, Whitfield, and Martin E. Hellman. Privacy and authentication: an introduction to cryptography. *Proceedings of the IEEE*, 67 (March 1979), 397–427.

[FIP77] FIPS Publication 46. Specifications for the Data Encryption Standard. U.S. Dept. of Commerce, National Bureau of Standards, (January 15, 1977).

[FIP80] FIPS Publication 81. DES modes of operation. U.S. Dept. of Commerce, National Bureau of Standards, (December 2, 1980).

[Gal68] Gallager, R. G. *Information Theory and Reliable Communication*, John Wiley, (New York, 1968).

[Gif82] Gifford, David K. Early experience with natural random bits. Seminar talk, MIT Laboratory for Computer Science, (May 11, 1982).

[GoM81] Goldwasser, Shafi, and Silvio Micali. A bit by bit secure public-key cryptosystem. Technical memo UCB/ERL M81/88, Univ. of California, Berkeley, (December 1981).

[GoM82] Goldwasser, Shafi, and Silvio Micali. Probabilistic encryption & how to play mental poker keeping all partial information secret. *Proceedings of the 14th Annual ACM Symposium on Theory of Computing*, (May 5-7, 1982), 365–377.

[Kah67] Kahn, David. *The Codebreakers: The Story of Secret Writing*, Macmillan, (New York, 1967).

[Kle60] Kleinrock, L. A program for testing sequences of random numbers. MIT Lincoln Laboratory Report 51G-0018, (October 25, 1960).

[Kru81] Kruh, Louis. The Genesis of the Jefferson/Bazeries Cipher Device. *Cryptologia*, 5 (October 1981), 193–208.

[Lem79] Lempel, Abraham. Cryptology in transition. *ACM Computing Surveys*, **11** (December 1979), 285–303.

[Mad72] Maddocks, R. S. *et al.* A compact and accurate generator for truly random binary digits. *Journal of Physics E: Scientific Instruments*, **5** (1972), 542-544.

[McE78] McEliece, R. J. A public-key cryptosystem based on algebraic coding theory. Deep Space Network Progress Report 42-22, Pasadena Jet Propulsion Labs., (January-February 1978), 114–116.

[Mer78] Merkle, Ralph C. Secure communications over insecure channels. *CACM*, **21** (April 1978), 294–299.

[Nic82] Nicolai, Carl R. Nondeterministic cryptography. *Advances in Cryptology: Proceedings of CRYPTO 82*, Plenum Press, (New York, 1983).

[NiZ80] Niven, Ivan, and H. S. Zuckerman. *An Introduction to the Theory of Numbers*, John Wiley, (New York, 1980).

[Rab78] Rabin, Michael O. Digitalized signatures. *Foundations of Secure Computation*, (edited by DeMillo *et al.*). Academic Press, (New York, 1978), 155–168.

[Rab79] Rabin, Michael O. Digitalized signatures and public-key functions as intractable as factorization. Technical report no. TR–212, MIT Lab. for Computer Science, (January 1979).

[Rab81] Rabin, Michael O. How to exchange secrets by oblivious transfer. Technical memo TR–81, Harvard Center for Research in Computing, (1981).

[SRA79] Shamir, Adi, Ronald Rivest, and Leonard Adleman. Mental poker. *The Mathematical Gardner* (edited by D. Klarner), Prindle, Weber, and Schmidt, (Boston, 1981), 37–43.

[Sha49] Shannon, Claude E. Communication theory of secrecy systems. *Bell System Technical Journal*, **28** (October 1949), 659–715.

[Sim82] Simmons, Gustavus J., and Diane Holdridge. Forward search as a cryptanalytic tool against a public key privacy channel. Presented at the Symposium on Computer Security and Privacy, (Oakland, April 1982).

[Slo82] Sloane, N. J. A. Error-correcting codes and cryptography—part I. *Cryptologia*, **6** (April 1982), 128–153.

[Wyn75] Wyner, A. D. The wire-tap channel. *The Bell System Technical Journal*, **54** (October 1975), 1355–1387.

Session III: Protocols and Transaction Security

ON THE SECURITY OF MULTI-PARTY PROTOCOLS IN DISTRIBUTED SYSTEMS

Danny Dolev[†]

Institute of Mathematics and Computer Science
Hebrew University, Jerusalem

Avi Wigderson

Electrical Engineering and Computer Science Department
Princeton University
Princeton, NJ 08544

ABSTRACT

Security of protocols for network communication has received considerable attention in recent years. We concentrate on ensuring the security of cryptographic protocols in distributed systems.

In a distributed system, beyond eavesdropping, a saboteur may impersonate another user or alter messages being sent. A saboteur who is also a user may send conflicting messages or use other illegal messages in order to uncover secret information.

The problem we address, in its most general form, is: "given a multi-party protocol which is provably secure when all the participants monitor every message being sent, can the protocol be modified to be secure in a distributed system?"

We use the Byzantine Agreement, Crusader Agreement, and other specific checks to improve protocols by making them secure in a general distributed network. We examine the trade-off between detection of faulty behaviour and the number of messages exchanged.

1. Introduction

The main purpose of a cryptosystem is to protect private information. Cryptographic protocols enable users to employ the cryptosystem while com-

---
[†] Part of this work has been done while the first author visited IBM Research Center, San Jose, California.

municating with each other. A cryptosystem is secure if it cannot be "broken" using various mathematical tools. The security of a cryptographic protocol is measured by its ability to prevent eavesdroppers and other non-participants from understanding the information being exchanged. Distributed cryptographic protocols are intended to guarantee the security of private information in case one cannot trust even the users participating in the protocol itself. We will make sure that private information will not be disclosed even in the case that several participants collaborate in attempting to cause such disclosure.

The security of cryptographic protocols has been discussed in previous papers (DEK, DLM, DY, LW, NS). In this paper we assume that we are given a protocol which is secure in a *round-table environment* (or round-table secure). That is, the protocol is secure provided every user is able to see every message being exchanged between every pair of users in the system, provided users are consistently following the protocol, and provided deviations from the protocol can be detected. We will show how to modify such a protocol to make it secure in a distributed system in which a user can see only information he himself receives. For example, all the protocols presented in (LW) are round-table secure. But the same protocols, when used in a distributed system are no longer secure.

The difference between a round-table and a distributed system is that in the latter a user can slightly deviate from the protocol in a way that the recipient of the message cannot be aware of, and by doing so he is able to uncover secret information. As a matter of fact, a saboteur in a distributed system can do can do whatever he wants as long as the user receiving the message does not suspect a misbehaviour, or even if he suspects, he is unable to obtain a "proof of misbehaviour".

The methods we will give for improving round-table secure protocols by making them distributed-secure produces protocols that are fault-tolerant in a very broad sense. The protocols will be able to sustain **any** malicious behaviour without ever revealing an individual's secret. We make no assumptions about the type of faulty behaviour nor about the communication network. We guarantee that in the case that the participants are all faithful, the protocol proceeds as usual. A malfunction may cause the protocol to stop. However, even in this case a saboteur cannot uncover private information. The algorithms we use will try to overcome the faulty behaviour, but we cannot guarantee to identify all the faulty users, as one might have hoped. The reason is that a faulty user can behave in a way that does not disclose his faultiness. However in cases where faultiness may endanger the security of the protocol, a faulty user will be detected.

2. The Model

Let U be a set of users participating in a given network. For simplicity we assume that all the users in U participate in the protocol. Moreover, we assume that the network is such that every two users can communicate directly. To relax this assumption one can follow the results of (Da, LSP) and obtain similar restrictions on the network connectivity. Let CU be the subset of correct (or faithful) users, and TU be the rest of the users, named the traitors (faulty users or saboteurs). We assume that faithful users do not know which users are members of which sets. Faithful users behave correctly; they follow the protocols and correctly execute the algorithms. There are no assumptions on the behaviour of saboteurs; they may even collaborate in trying to break the protocol. Notice that saboteurs may know who are the members of each set. Moreover, the can behave faithfully, or at least pretend to do so. We do assume that saboteurs cannot break the cryptosystem.

Let t be an upper bound on the number of saboteurs that may participate in the protocol, that is, t is the cardinality of the set TU. Our results are parametrized by t.

We assume that users communicate synchronously and thus an absence of a message can be detected. The synchronous behaviour can be relaxed, but then one needs an upper bound for the time it takes for a faithful user to respond to a message he receives (FLP).

In the analysis that follows, we assume that no message has been stopped by eavesdroppers. To overcome the stopping we need to include this misbehaviour in the parameter t in some way (if this possibility was not limited, one faulty user could have isolated every other user from all others). We will not cover this extension here.

Let MSG be the set of possible messages. We assume MSG is closed under the operation of forming sequences from MSG.

A *synchronized multi-party protocol* P *for a set of users* U is a finite sequence of phases $G_1, G_2, \cdots, G_l$. Each phase is a directed graph $G_k(U, E_k, T_k)$ where nodes correspond to the users in U, and the edges E_k are labeled from MSG according to T_k. For every directed edge (u,v) in E_k, $T_k(u,v) = f(k, I_u, M(k,u))$, where f is a function of I_u, the private information of user u, and $M(k,u)$, that contains the information u obtains through phase k from messages he receives. ($M(1,u)$ is empty for every faithful user u).

The phases are numbered with consecutive positive integers. At each phase k, a user should send the messages on his outedges and receive the messages on his inedges.

A protocol works in a *round-table* environment if for every faithful user u, $M(k,u) = M(k-1,u) \cup \{T_k(a,b) \mid a,b \in U\}$. In other words, every faithful user has the same information about every message that was sent in each phase.

A protocol is *round-table secure* if it is secure under the condition that a faithful user sends a message at the kth phase only if for every pair of faithful users a,b we have $M(k-1,a) = M(k-1,b)$.

A protocol works in a distributed system if at every phase k and for every faithful user u, $M(k,u) = M(k-1,u) \cup \{T_k(a,u) \mid a \in U\}$. In other words, a faithful user obtains only the messages he received at each phase.

We assume the exsistence of a cryptographic signature scheme with the following properties:

(1) Every user x has a distinct signature function S_x with which he signs messages.
(2) Every user can identify the signature of every other user and can extract the message from the signed message.
(3) No user (not even a saboteur) can forge the signature of another user.
(4) Every change in a signed message which is not done by the user who signed it can be detected by any other user.
(5) The signature functions do not commute, i.e. for every message M $S_x S_y(M) \neq S_y S_x(M)$ whenever $x \neq y$.

A signature scheme with the above properties can be constructed using a (secure) public-key cryptosystem (DH, RSA). We further assume that the cryptographic system used for signatures is independent of the one used for encryption.

using the cryptographic signature scheme one can obtain messages with several signatures, can extract the contents of a message, and can identify the various signatures (and their order) a message carries. This will be the way we will use the signature scheme in the coming sections.

We use the following notation: A user notarizes a message if he signs it and sends it to all users. (Our algorithm works also if faulty users are able to remove signatures notarizing a message in an undetectable way). For simplicity we will assume that when a faithful user sends a message he also sends it to himself. A message is said to be k-dense if its contents is followed (notarized) by exactly k distinct signatures. Notice that the contents of a message can be another message which by itself carries signatures. We assume that the content of a message is uniquely defined. (In fact, it follows from our definition of a protocol).

3. Distributed Agreements

The security of a round-table secure protocols is based on the fact that all the faithful users obtain the same information about every message being exchanged in each phase. The idea is that by obtaining that information they can check and find deviations from the protocol and stop the protocol before the saboteur is able to break it. We will try to bring the users in a distributed system to an agreement on what messages each user has sent. By doing that we will induce the security of a protocol in a distributed system from its security in a round-table environment. Our exact goal is : "Given a round-table secure protocol P, produce a modified distributed protocol P' such that at every phase of P' which correspond to a phase of P there is an agreement about the messages sent in the previous phase of P, or at the first time that this does not hold, at least one faithful user holds a "proof" about the faultiness of some user".

Two types of distributed agreements have been discussed previously in the literature: Byzantine agreement and Crusader agreement. These two agreements will enable us to improve a round-table secure protocol. For the Crusader agreement let us assume that there exists a transmitter who is supposed to send his value to the network and that all the users know when he is supposed to do so.

Crusader Agreement (Da)
(1) If the transmitter is faithful, then all the faithful users should agree on the value he has sent.
(2) All the faithful users who do not hold a "proof" of faultiness of the transmitter should agree on the same value.

The elements of the Crusader agreement are that as long as all users are faithful they reach agreement about every value being sent. When faultiness of the transmitter is introduced then the faithful users are divided into two sets (the users do not necessarily know who are the members of each set). One set includes all the faithful users holding a "proof" of the faultiness of the transmitter, and the second set is the set of all the faithful users who do not hold such a proof. All members of the second set should decide on the same value as the value of the transmitter.

In the following algorithm a notarized message carrying a value of the transmitter should contain also the transmitter's signature.

The Crusader Algorithm:
C1. The transmitter notarizes his value at phase 1.
C2. If at the end of phase 1 a single value signed by the transmitter is received, then notarize it at the next phase. Otherwise notarize the message "the transmitter is faulty".
C3. If at the end of phase 2 at least 2 2-dense messages containing different values or at least $t+1$ signed messages saying that the transmitter is faulty are received define it to be a *proof of faultiness*. Otherwise agree on the only notarized value you have received.

Theorem 1:
If the cardinality of the set of faulty users is bounded by t, then the Crusader algorithm reaches the Crusader agreement at the end of phase 2.

The proof is similar to the proof in (Da). Observe that the proof of faultiness a faithful user can hold at the end of the algorithm is either two different values signed also by the transmitter or $t+1$ signed messages claiming that the transmitter is faulty. Neither of these can be produced unless the transmitter itself is faulty.

Lemma 1:
A faithful user holds a proof of faultiness only if the transmitter is faulty. If the transmitter is faithful, then no user (even a faulty one) can hold a proof of faultiness.

The natural idea is to run the algorithm to obtain the Crusader agreement about every message $T_k(u,v)$ being sent and by that to reach some sort of agreement among the faithful users about the various messages. The problem is that at the end of the Crusader agreement some of the faithful users do agree on the message but some do not. So if those who did not find a "proof" of faultiness continue to follow the protocol they will send the next messages when the conditions about $M(k,u)$ do not hold.

We can obtain a round-table environment by running a Crusader agreement on every message and then adding another phase in which only users who hold a proof of faultiness will send it and the rest will wait a phase without doing anything. By doing this we ensure that every user who does not receive a proof of faultiness by the end of that phase can be sure that all the faithful users have obtained agreement on the value of the transmitter at the end of the Crusader agreement. Therefore if he continues that protocol he cannot cause any security problem because of the fact that the round-table conditions held at the previous phase.

If a faithful user obtains a proof of faultiness at the end of the Crusader algorithm, then by the end of the additional phase all the faithful users will learn about that. This will prevent them from continuing the protocol and will stop the protocol at that point without enabling the faulty users to break it.

Observe that it still may happen that some of the faithful users will see a proof of faultiness at the end of the additional phase and will stop taking part in the protocol because of that. This will bring about a situation in which not all the users have the same information. But in this case those that will continue will be covered because of the round-table condition and those that have stopped will try to stop the protocol before further phases will be completed.

To do this we actually need to run another algorithm that will distribute that proof of faultiness among the faithful users in order to bring them to an agreement about the faultiness of the transmitter. For that we use the Byzantine agreement. For the Byzantine agreement let's assume that users may hold a legal proof (of faultiness) and that every user can check its legality.

Byzantine Agreement (Da, DS, LSP, PSL)
(1) If a faithful user holds a legal proof, then all the faithful users should agree on the existence of a legal proof.
(2) All the faithful users should reach the same agreement.

The difference between the two agreements seems to be small but it is important. The Byzantine agreement requires reaching agreement no matter what. The Byzantine agreement requires more phases than the Crusader agreement. We will run it after completing the Crusader agreement in parallel with the rest of the protocol in such a way that if a fault has been found, the Byzantine agreement will broadcast it and will stop the rest of the protocol.

The Byzantine agreement will be used here to agree on the faultiness of some user. For this reason the version of the algorithm presented here will be optimized for this use.

The Byzantine Algorithm:
B1. Every faithful user who holds a legal proof notarizes his proof at phase 1.

B2. If at the end of phase $k < t+1$ a k-dense legal proof of faultiness is received by a user who did not notarize it before, then he notarizes it at the next phase and agrees on the proof of faultiness.

B3. If by the end of phase $t+1$ a $t+1$-dense proof of faultiness is received, then agree on the proof of faultiness.

Theorem 2:

If the cardinality of the set of faulty users is bounded by t, then the Byzantine algorithm reaches the Byzantine agreement at the end of phase $t+1$. If at some phase $k < t+1$ some faithful user agrees on a legal proof of faultiness, then by phase $k+1$ all the faithful users will agree on a legal proof of faultiness.

The proof is similar to the proof in (DS) although our algorithm and definition of Byzantine agreement is somewhat different.

4. Algorithm for Improving Security

Let P be a given protocol. The following algorithm will show how to obtain from it a protocol P' with the property that if P is round-table secure, then P' is secure for a distributed system. Furtheremore, P' will have three times as many phases as P, s.t. each phase k in P corresponds to phases $3k-2$, $3k-1$, $3k$ in P'.

A proof of faultiness is called legal if

(1) it is about sending or not sending some $T_k(a,b)$ that had to be sent by a at phase $3k-2$.

(2) it contains either two different values being sent (and signed) by the faulty user a at that phase or $t+1$ signatures of users who are supposed to receive $T_k(a,b)$ claiming not to receive it at that phase.

(3) it is k-dense.

(4) it is received by phase $k' < 3k$ of P'.

The legality of a given proof of faultiness can be checked, due to our assumptions, by every user who receives it. If a proof of faultiness is not legal, then the receiving user can ignore it.

To simplify the arguments in the following algorithm we assume that every message a user has to send according to protocol P contains the name of the user who suppose to receive that message in P.

Crusader-Byzantine algorithm

Define P' to be the protocol obtained from P as follows: for every $k > 0$

CB1. At phase 3k-2: for every faithful user u, if u did not agree on a legal proof of faultiness by now, then for every $T_k(u,v)$ that u is supposed to send according to the protocol P at phase k he uses the Crusader algorithm to send the message "$T_k(u,v)$, the phase is $3k-2$". (Recall that the Crsader algorithm takes two phases, $3k-2$ and $3k-1$).

CB2. At phase 3k: If a faithful user u holds a proof of faultiness about some user b at phase $3k-2$, then he starts a Byzantine agreement to send that legal proof of faultiness to all the users. Otherwise, he defines $M(k,u) = M(k-1,u) \cup \{T_k(a,b) \mid u$ has agreed on at phase $3k-1\}$.

CB3. If $3k-2$ is the earliest phase about which a faithful user u holds a legal proof of faultiness, then u "STOP"s the protocol at phase $3k+t+1$.

Lemma 2:
In the Crusader-Byzantine algorithm, if the cardinality of the set TU is bounded by t and a faithful user holds a legal proof of faultiness at some phase k, then by phase $k+1$ all the faithful users will hold a proof of faultiness. Moreover, if a legal proof of faultiness about some phase $3r-2$ has not been received by any faithful user by phase $3r+t$, then no faithful user will ever later accept any proof of faultiness about phase $3r-2$ as being legal.

The proof follows from the properties of the Byzantine agreements and the way we use them in the Crusader-Byzantine algorithm.

Notice that we can save one phase in lemma 2 and in stopping the protocol if we would exclude the faulty user while running the Byzantine agreement about his fault.

Lemma 3:
In the Crusader-Byzantine algorithm, if a faithful user u decides to send $T_k(u,v)$ at phase $3k-2$ of P, then $M(k-1,a)=M(k-1,b)$ for every pair a,b of users.

These lemmas enable us to prove the main theorem about the security of the new protocol P'.

Theorem 3:
If the protocol P is round-table secure and the cardinality of TU is bounded by t, then the protocol P' obtained from P using the Crusader-Byzantine algorithm is secure. Moreover, in case the protocol P' stops with a proof of faultiness, then a faulty user is found and this fact is known to all the users.

Observe that if the security of P holds only in the case where t is bounded by some function of $n=|U|$, then in the new protocol P' it will also be bounded by that function of n. For example the round-table secure protocols in (LW) require that $t<n-1$ or $t<n/2$, and therefore in the protocol one obtains from it using the Crusader-Byzantine algorithm, t should also be be similarly bounded for ensuring the security of the protocol.

5. Conclusions

The number of phases in the new protocol, P', is only three times as many as the number of phases in P. However, The number of messages that P' uses is more than that of P by a factor of n^2. Using better algorithms for reaching Byzantine agreement and similar ideas in the Crusader agreement one can reduce this factor to nt. One can save even more, using the algorithm in (DR), but that algorithm will require many more phases for reaching the agreement about the faultiness of a user. Observe that if one is not interested in reaching agreement about the faultiness of a user, then the number of messages can be further reduced, and so can the number of phases.

The algorithm we presented gives a way to induce security of distributed protocols from that of a round-table protocol. Thus, a protocol designer can concentrate on producing round-table secure protocols, and then convert them using our algorithm to be secure in a distributed system. Further research is needed for obtaining the most efficient way to transform nondistributed secure protocols to be secure in a distributed system.

The ideas presented in the paper can also be used for networks in which not every two users can communicate directly. In addition, the ideas can be used for improving protocols that are secure in environments fulfilling weaker assumptions than the round-table environment.

References

(Da) D. Dolev, "The Byzantine Generals Strike Again," Journal of Algorithms, vol. 3, no. 1, pp.14-30, 1982.

(DEK) D. Dolev, S. Even, and R. M. Karp, "On the Security of Ping-Pong Protocols," CRYPT82, Santa Barbara, Aug. 1982.

(DH) W. Diffie, and M. Hellman, "New Direction in Cryptography," IEEE Trans. on Information Theory, IT-22, 6, pp. 644-654, 1976.

(DLM) R. A. DeMillo, N. A. Lynch, and M. Merritt, "Cryptographic Protocols," Proceedings of the 14th ACM SIGACT Symposium on Theory of Computing, May 1982.

(DR) D. Dolev, and R. Reischuk, "Bounds on Information Exchange for Byzantine Agreement," Proceedings of the ACM SIGACT-SIGOPS Symposium on Principles of Distributed Systems, Aug. 1982.

(DS) D. Dolev, and H. R. Strong, "Polynomial Algorithms for Multiple Processor Agreement," Proceedings of the 14th ACM SIGACT Symposium on Theory of Computing, May 1982.

(DY) D. Dolev, A. C. Yao, "On the Security of Public Key Protocols," to appear, IEEE Trans. on Information Theory.

(FLP) M. J. Fischer, N. A. Lynch, and M. S. Paterson, "Impossibility of Distributed Consensus with One Faulty Process," unpublished manuscript, Aug. 1982.

(LSP) L. Lamport, R. Shostak, and M. Pease, "The Byzantine Generals Problem," ACM Trans. on Programming Languages and Systems, to appear.

(LW) R. J. Lipton, and A. Wigderson, "Multi-Party Cryptographic Protocols," unpublished manuscript, May 1982.

(NS) R. M. Needham, and M. D. Schroeder, Using Encryption for Authentication in Large Networks of Computers," CACM, vol.21, no. 12, pp. 993-999, 1978.

(PSL) Presence of Faults," JACM, vol. 27, no. 2, pp. 228-234, 1980.

(RSA) R. Rivest, A. Shamir, and L. Adleman, "A method for obtaining digital signatures and public-key cryptosystems," CACM, vol. 21, pp. 120-126, 1978.

ON THE SECURITY OF PING-PONG PROTOCOLS

(Extended Abstract)

D. Dolev	S. Even	R. M. Karp
IBM Research Lab San Jose, CA 95193	Computer Science Technion Haifa, Israel	Computer Science University of CA Berkeley, CA 94720

ABSTRACT: Consider the class of protocols, for two participants, in which the initiator applies a sequence of operators to a message M and sends it to the other participant; in each step, one of the participants applies a sequence of operators to the message received last, and sends it back. This "ping-pong" action continues several times, using sequences of operators as specified by the protocol. The set of operators may include public-key encryptions and decryptions.

We present an $O(n^3)$ algorithm which determines the security of a given protocol (of length n). This is an improvement of the algorithm of Dolev and Yao [DY].

I. INTRODUCTION

The use of public-key encryption [DH,RSA] for secure network communication has received considerable attention. Such systems are effective against a "passive" eavesdropper, namely, one who merely taps the communication line and tries to decipher the intercepted message. However, as pointed out by Needham and Schroeder [NS], an improperly designed protocol can be vulnerable to "active" sabotage.

The saboteur may be a legitimate user in the network. He can intercept and alter messages, impersonate other users or initiate instances of the protocol between himself and other users, in order to use their responses. It is possible that through such complex manipulations he can read messages, which are supposed to be protected, without cracking the cryptographic systems in use.

In view of this danger it is desirable to have a formal model for discussing security issues in a precise manner, and to investigate the existence of efficient algorithms for checking the security of protocols.

Dolev and Yao [DY] investigated the security of what we call here "ping-pong protocols." These protocols involve two participants, the sender S and the receiver R. Let M be a message generated by S. First, S applies a sequence of operators to M and sends it to R. Next, R applies a sequence of operators to the message received, and sends the result back to R. In each step, the participant applies a sequence of operators to the last message received, and sends it back. The number of times this is done, as well as the sequences of operators used, is defined by the protocol.

Dolev and Yao considered the security of two such families of protocols, assuming only few limitations on the behavior of the saboteur. Their second and more general family of protocols is extended here to allow more operators, and an $O(n^3)$ time algorithm for checking the security of protocols is presented. This improves the algorithm of Dolev and Yao, which is $O(n^8)$ time.

We briefly recall the essence of public-key systems (see [DH] or [RSA] for more details). Every user X has an encryption function E_X and a decryption function D_X. Both are mapping from $\{0,1\}^*$ into $\{0,1\}^*$. There is a public directory containing all (X,E_X) pairs, while the decryption function D_X is known only to X. The main requirements on E_X, D_X are:
    (1) $E_X D_X = D_X E_X = \lambda$, where λ is the identity function, and
    (2)   Knowledge of $E_X(M)$ does not reveal anything about the value M.

Before we attempt any formal definitions of protocols or security let us consider several simple examples of protocols, and discuss informally their security.

Example 1. Consider the following protocol:
    (1) $(X, E_Y(M), Y)$
    (2) $(Y, E_X(M), X)$
which simply means this: X wants to send M to Y and gets an echo in order to verify that M has reached Y. He computes $E_Y(M)$, using Y's public encryption key and sends via the network $(X, E_Y(M), Y)$, which stands for "X sends to Y the message $E_Y(M)$." Clearly, no one but Y can apply D_Y to

$E_Y(M)$, in order to recover M. After doing so, Y computes $E_X(M)$ and sends $(Y,E_X(M),X)$. When X gets it he can compare the echo, $D_X E_X(M)$, with the original M in order to verify that M has indeed reached Y.

This innocent-looking protocol is insecure. A saboteur Z may intercept $(X,E_Y(M),Y)$ and replace it by $(Z,E_Y(M),Y)$. Y will get M, and respond, according to the protocol by sending $(Y,E_Z(M),Z)$. Z can now read M by applying his secret key, D_Z. He can then even produce the echo $(Y,E_X(M),X)$ and send it over to the satisfied and unsuspecting X. Clearly, this works only if M itself does not include information about the original sender's identity. Indeed, this observation leads to the technique of name-appending:

Example 2.
    (1) $(X,E_Y(MX),Y)$
    (2) $(Y,E_X(M),X)$

The word MX is formed by appending to M the name X. Now, after Y applies D_Y to $E_Y(MX)$ to get MX, he checks whether the suffix of the string matches the declared name of the sender, i.e., X. If it does not, he knows that someone has meddled with the message and simply terminates his participation in this instance of the protocol. Otherwise, he computes $E_X(M)$ and sends $(Y,E_X(M),X)$.

This protocol is indeed secure. A formal way to prove it will be shown in Section III.

One may be led to believe that name-appending is the cure to all evils, but consider this seemingly "even safer" protocol.

Example 3.
    (1) $(X,E_Y(E_Y(M)X),Y)$
    (2) $(Y,E_X(M),X)$

Here the X is appended to $E_Y(M)$ instead of to M itself. This protocol is insecure!

In addition to the operator E_X and D_X, we have used in the last two examples two more operators, which we shall denote i_X and d_X. If X is a string (name of user X) and M is a string (message) then $i_X(M)=MX$. Let S be a string, $d_X(S)$ is defined as follows. If X is a suffix of S, i.e., $S=MX$, then $d_X(S)=M$; else, $d_X(S)$ is undefined, which means that the participation in this instance of the protocol is terminated. Clearly,

$$d_X i_X = \lambda$$

but $i_X d_X(S)$ is not even defined (unless $S=MX$ for some M).

Let us also define an operator d, which is simply the removal of the appended user name. This is easy to do, if, for example, all names use exactly the same number of bits. Therefore, it is natural to assume that a saboteur can perform d. Thus, for every user name X,

$$di_X = \lambda$$

but again $i_X d(S) = S$ only if $S = MX$.

In general we shall assume that there is a set of operators, Σ, which can be used by the participants in the network. Some operators may have a user name subscript (such as E_X, D_X, i_X and d_X in our examples). The subset of operators, which user X can perform will be denoted by Σ_X and will be called X's <u>vocabulary</u>. The vocabularies of all users are similar in the sense that if one replaces the index X by Y, and Y by X, in Σ_X, the result is Σ_Y.

Also, there will be a given set of <u>cancellation rules</u> of the form $\sigma\tau = \lambda$, where σ and τ are elements of λ. If both σ and τ are indexed then the indices are the same. The cancellation rules are similar for all users. Thus, if one or both operators are indexed then the same cancellation rule holds for every user name index.

In our examples,

$$\Sigma = \{d\} \cup \{E_X, D_X, i_X, d_X \mid X \text{ is a user name}\}$$

$$\Sigma_X = \{d, D_X\} \cup \{E_Y, i_Y, d_Y \mid Y \text{ is a user name}\},$$
and the cancellation rules are

$$E_X D_X = \lambda$$

$$D_X E_X = \lambda$$

$$d_X i_X = \lambda, \text{ and}$$

$$di_X = \lambda.$$

Note that if $a, b, c \varepsilon \Sigma$, $ab = \lambda$ and $bc = \lambda$ then $a = c$. This follows from the fact that members of Σ are operators: Let $w \varepsilon \{0,1\}^*$.

abc(w) = a(bc(w)) = a(w), since bc=λ, but on the other hand
abc(w) = ab(c(w)) = c(w).
Thus a=c.

Given a string $\alpha \varepsilon \Sigma^*$, one may repeatedly apply cancellation rules until no cancellation rule is applicable any more. By previous paragraph, the reduction process has the Church Rosser property [R], and thus the end result is unique. Let us denote this reduced form of α by $\bar{\alpha}$.

An underlying assumption in our analysis is that the set Σ is free from any relations other than those implied by the cancellation rules. That is, two strings of operators, α and β, are equivalent if and only if both have the same reduced form.

II. PING-PONG PROTOCOLS AND SECURITY

<u>Definition</u>. A ping-pong protocol P(S,R) is a sequence $\Gamma = (\alpha_1, \alpha_2, ..., \alpha_\ell)$ of operator-words, such that if i is odd then $\alpha_i \varepsilon \Sigma_S^*$ and if it is even then $\alpha_i \varepsilon \Sigma_R^*$.

The structure of the protocol is similar for every ordered pair of (different) users. Thus, if in P(V,W), we replace the index V by X, and W by Y, we get P(X,Y). We assume that for every two users X and Y, P(X,Y) may be initiated, i.e., there are no restrictions, imposed by the network or the users, on communication via P.

In Example 1, $\alpha_1[S,R] = E_R$, $\alpha_2[S,R] = E_S D_R$ and $\ell = 2$.

In Example 2, $\alpha_1[S,R] = E_R i_S$, $\alpha_2[S,R] = E_S d_S D_R$ and again $\ell = 2$. In both examples, $\alpha_1(M)$ is sent by S to R and $\alpha_2 \alpha_1(M)$ is sent by R, back to S.

In general the interpretation is as follows: S invents a message-word $M \varepsilon \{0,1\}^*$. He applies α_1 to it and sends it to R; i.e., the first step is $(S, \alpha_1(M), R)$.

Next $(R, \alpha_2 \alpha_1(M), S)$, etc.

If ℓ is odd the last step is $(S, \alpha_\ell \alpha_{\ell-1} ... \alpha_1(M), R)$ and if ℓ is even, then it is $(R, \alpha_\ell \alpha_{\ell-1} ... \alpha_1(M), S)$.

In this paper, we are not concerned with the purpose of using P. Instead, we are interested only in the question of whether a saboteur (or a group of them) can extract M.

Thus, we assume that some user, S, has invented a message M, chosen a user R and initiated P(S,R) on M. We assume that neither S, nor R, is a saboteur. We have to define what are the actions which the saboteur(s) can take.

We shall assume that for every $1 \leq i \leq \ell$, for every two different users X and Y and for every $W \varepsilon \{0,1\}^*$ the saboteur can effect $\alpha_i[X,Y]$ on W.

We shall explain, shortly, why we make this assumption, but if one believes that this is too conservative one may restrict the saboteur actions, and as long as these restrictions are symmetric (not user-name dependent), an $O(n^3)$ algorithm for checking security still exists. For example, one could assume that a saboteur cannot effect $\alpha_1[S,R]$ if he is not S, or that he cannot effect $\alpha_i[S,R]$, $1 \leq i \leq \ell$, if he is not S.

Let us denote the saboteur by Z.

If X=Z, then Z has no difficulty to get $\alpha_1(W)$, since $\alpha_1 \varepsilon \Sigma_Z^*$. If X$\neq$Z (and X is not one of the collaborating saboteurs) Z may be able to convince X to initiate P(X,Y) on W. By tapping the message $(X,\alpha_1(W),Y)$, Z will get $\alpha_1(W)$.

In order to effect $\alpha_i[X,Y]$ on W, for $1 \leq i \leq \ell$, Z can wait for P(X,Y) to occur (or somehow convince X to initiate it), wait for the (i-1)st message, $(X,\alpha_{i-1},\alpha_{i-2}...\alpha_1,Y)$--assuming i is even, intercept it and replace it with (X,W,Y). Now, Y responds with $(Y,\alpha_i(W),X)$, as expected of him (assuming $\alpha_i(W)$ is defined) and sends it through the network, where Z can tap it.

It follows that the language of operator-words which a single saboteur Z can effect (on any $W \varepsilon \{0,1\}^*$) is

$$\Delta = [\Sigma_Z \cup \{\alpha_i[X,Y] \mid 1 \leq i \leq \ell, X \text{ and } Y \text{ are different users}\}]^*.$$

<u>Definition</u>. Let $\alpha_1[S,R]$ be the first operator-word of P(S,R) and $Z \notin \{S,R\}$. P is <u>insecure</u> if there exists an operator-word $\gamma \varepsilon \Delta$ such that $\overline{\gamma \alpha_1} = \lambda$.

Observe that it is not necessary to consider $\alpha_i \alpha_{i-1}...\alpha_1(M)$, for $1 \leq i \leq \ell$, which is also heard over the network. For if a $\gamma \varepsilon \Delta$ exists which satisfies $\overline{\gamma \alpha_i \alpha_{i-1}...\alpha_1} = \lambda$ then there is a $\gamma' \varepsilon \Delta$ (in fact $y' = \gamma \alpha_i \alpha_{i-1}...\alpha_2$ will do) for which $\overline{\gamma' \alpha_1} = \lambda$.

In the definition of security given above, we called P insecure if for some ordered pair of users (not including saboteurs), a $\gamma \varepsilon \Delta$ exists for which $\overline{\gamma \alpha_1} = \lambda$. In fact, such a γ exists for one pair (S,R) if and only if it exists for every set of users. This follows immediately from the fact that change of names of users does not change the pattern of cancellations. Thus, in what follows we shall restrict our attention to a fixed pair of users, (S,R), free of saboteurs, and <u>only</u> consider the question of whether for $\alpha_1[S,R]$ a $\gamma \varepsilon \Delta$ exists which satisfies $\overline{\gamma \alpha_1} = \lambda$.

One may wonder why we have defined Δ to include Σ_Z, but have not allowed a set of saboteurs $\{Z_1, Z_2, \ldots Z_m\}$ and put

$$\bigcup_{i=1}^{m} \Sigma_{Z_i}$$

in Δ instead. It can be shown that this is not necessary, since whatever a set of saboteurs can do, a single saboteur can do also.

Our next goal is to restrict Δ even further, in order to simplify the security decision problem. Let us show that if a $\gamma \varepsilon \Delta$ exists, for which $\gamma \cdot \alpha_1[S,R] = \lambda$, then the same statement holds for the following Δ':

$$\Delta' = \{\Sigma_Z \cup \{\alpha_i[X,Y] \mid 1 \le i \le \ell, X=Y \text{ and } \{X,Y\} \subset \{R,S,Z\}\}\}^*.$$

If we replace each user $U \notin \{R,S\}$ who appears in γ by Z, the cancellation pattern is maintained while each α_i either remains legitimate (with two different users $X,Y, \{X,Y\} \subset \{R,S,Z\}$) or becomes an operator-word in Σ_Z^*. This proves that we can replace Δ by Δ' in the security decision problem.

III. AN ALGORITHM FOR CHECKING PROTOCOL SECURITY

Construct a nondeterministic finite state automaton A, as follows:

(1) State 0 is the (unique) initial state and state 1 is the (unique) accepting state. The (input) alphabet is $\Sigma = \Sigma_Z \cup \Sigma_S \cup \Sigma_R$.

(2) There is a direct path from state 0 to state 1 whose (input) labels correspond to $\alpha_1[S,R]$.

(3) For every input letter (operator) $\sigma \varepsilon \Sigma_Z$, there is a self-loop from 0 to 0, labelled σ.

(4) For every $\alpha_i[X,Y]$, $1 \le i \le \ell$ and $\{X,Y\} \subset \{R,S,Z\}$ there is a loop from 0 to 0 whose edges are labelled, in sequence, by the letters of α_i.

Let us assume that the (simplified) automaton A has been constructed and that its set of states is $S = \{0, 1, \ldots, s\}$.

We say that a directed path, p, in A <u>collapses</u>, if its corresponding word w collapses, i.e., $\overline{w} = \lambda$.

Define the <u>collapsing relation</u> $C \subseteq S \times S$ as follows: $(i,j) \varepsilon C$ if there is a directed path from i to j, in A, which collapses.

The security question is therefore reduced to the question of whether $(0,1) \varepsilon C$. The protocol P is secure if and only if $(0,1) \notin C$.

In what follows, $i \xrightarrow{\sigma} j$ stands for an edge from state i to state j, labelled σ. Q is a queue of pairs of states. Our algorithm for constructing C is as follows:

(0) $C \leftarrow \{(i,i) \mid 0 \leq i \leq s\}$, $Q \leftarrow C$. [<u>Comment</u>: Each new pair of C enters Q once]
<u>while</u> $Q \neq \emptyset$, <u>do</u>

(1)   Delete the first pair, (i,j), from Q.

(2)   <u>If</u> $(j,k) \varepsilon C$ and $(i,k) \notin C$ <u>then</u> put (i,k) in C and in Q.

(3)   <u>If</u> $(k,i) \varepsilon C$ and $(k,j) \notin C$ <u>then</u> put (k,j) in C and in Q.

(4)   <u>If</u> $k \xrightarrow{\sigma} i$ and $j \xrightarrow{\tau} \ell$ and $\sigma\tau = \lambda$ [is one of the cancellation rules] and $(k,\ell) \notin C$ <u>then</u> put (k,ℓ) in C and in Q. <u>od</u>

The algorithm terminates, since there can be at most $(s+1)^2$ pairs in C and each can cause the loop to occur once; the number of operations in each pass of steps (1) through (4) is bounded by $O(s^3 + |\Sigma|^2)$. We shall shortly examine the time complexity questions more closely.

<u>Theorem 1</u>: The algorithm generates the collapsing relation C of automaton A. □

In the complexity analysis which follows, we assume the RAM model, and that the basic word-length is sufficient to accommodate all the operators. Thus, the test of whether $\sigma\tau = \lambda$ takes constant time.

<u>Theorem 2</u>: The time-complexity of the algorithm for constructing the collapsing relation of automaton A (of s+1 states) is $O(s^3 + s|\Sigma_Z|)$.

Let us denote by n the <u>length of the protocol</u> P, which is measured as follows:

$$n = \sum_{i=1}^{\ell} |\alpha_i|,$$

where $|\alpha_i|$ is the length of the operator-word α_i. Thus, n is the total number of operators used in P. Since each word $\alpha_i[X,Y]$ generates exactly 6 loops in the automaton A, (one for each choice of an ordered pair of users (X,Y) out of the set {S,R,Z}), the number of states, s, of A is O(n), while the number of self-loops is $|\Sigma_Z|$. If the operators (and cancellation rules) are fixed and are not part of the input of the security problem, then $|\Sigma_Z|$ and the table of cancellation rules is of constant size.

Thus, Theorem 2 implies, immediately, the following corollary:

<u>Corollary 1</u>: For fixed vocabulary and cancellation rules, there exists a security checking algorithm of ping-pong protocols (of two users). Its time-complexity is $O(n^3)$, where n is the length of the protocol.

In fact, one may also the definition of the generic vocabulary and cancellation rules to be part of the input, and still maintain the $O(n^3)$ bound on the time-complexity. One only needs to incorporate the preparation of the cancellation rules in form of a table into the algorithm (in time $O(n^2)$). Thus,

<u>Corollary 2</u>: For ping-pong protocols of two users there exists a security checking algorithm whose input is the generic cancellation rules and the protocol. Its time-complexity is $O(n^3)$, where n is the length of the input.

EPILOGUE

Essentially, the problem we have solved in Section III is that of checking whether the intersection of a regular language and a certain context-free language is nonempty. Classically, if one is given a context-free language L, by a grammar G in CNF, and a regular language R, by a nondeterministic automaton A, one constructs a new grammar G' which defines L∩R, and then one can check in linear-time whether G' defines the empty language. If the description of G is of length m and A is of n states then G' comes out of size $O(n^3m)$. Thus, this leads to an $O(n^3)$-time, $O(n^3)$-space algorithm to solve the security problem, while our solution is $O(n^3)$-time, $O(n^2)$-space. In fact, our algorithm can be generalized to answer the question of whether L∪R is empty, in $O(n^3m)$-time $O(n^2m)$-space.

Another issue is that of protocols for k>2 users. If one assumes that for $P(U_1,U_2,...,U_k)$ the saboteur can effect every α_i for k users, not necessarily distinct, then one saboteur is as powerful as many, and an $O(n^3)$ security checking algorithm similar to the one shown in Section III follows. However, it is natural to assume that this is not the case, since the user who is supposed to perform α_i, observing that not all k users are distinct, will become suspicious and will not cooperate.

Even the Goldreich have recently shown that there is an $O(k)$ bound on the number of "useful" saboteurs. Thus, for a fixed k an $O(n^3)$ security checking algorithm exists. However, if the number of users of P is part of the problem's input this observation is not useful since the straightforward extension of the algorithm leads to an exponential blow up. In fact, they show that if the cancellation rules are extended slightly (to allow commutativity of some operators) the security problem becomes undecidable.

The problem of testing the security of protocols which are not of the ping-pong type remains wide open.

ACKNOWLEDGMENT

The authors would like to thank Oded Goldreich and Michael A. Harrison for helpful discussions.

REFERENCES

[DH] W. Diffie and M. E. Hellman, "New Directions in Cryptography," IEEE Trans. Infor. Th., Vol. IT-22, No. 6, November 1976, pp. 644-654.

[RSA] R. L. Rivest, A. Shamir and L. Adleman, "A Method for Obtaining Digital Signatures and Public-Key Cryptosystems," Comm. ACM, Vol. 21, February 1978, pp. 120-126.

[NS] R. M. Needham and M. D. Schroeder, "Using Encryption for Authentication in Large Networks of Computers," Comm. ACM, Vol. 21, No. 12, December 1978, pp. 993-999.

[DY] D. Dolev and A. C. Yao, "On the Security of Public Key Protocols," to appear, IEEE Trans. on IT.

[R] B. K. Rosen, "Tree-Manipulating Systems and Church Rosser Theorems," JACM, Vol. 20, No. 1, January 1973, pp. 160-187.

THE USE OF PUBLIC-KEY CRYPTOGRAPHY FOR SIGNING CHECKS

Luc Longpré*

Computer Science Department
Cornell University
Ithaca, N.Y.

THE PROBLEM AND CONSTRAINTS

We want to build a secure system in which customers of a bank can make transactions and be able to keep a proof of each transaction. We also want the system to satisfy (as far as possible) the following constraints (our ultimate purpose is to remove all physical money):

- the customers are able to make transactions over the phone (just by exchanging messages),

- no communication with the bank is required for a transaction,

- no directory of customers is available.

By secure, we mean that nobody can use or destroy the money of a customer or use fake money. Secure also means that nobody can make a transaction without having the funds. We will also have to give a definition of a proof. Such a system is provably impossible with a usual definition of proof [1]. With a suitable (and reasonable) definition of proof, we will be able to use public-key cryptography.

The well-known checking system, besides the fact that it's not possible to make a transaction over the phone, is not secure for two reasons:

---

*Supported in part by a grant from NSERC

- it's relatively easy to imitate a signature,

- it's possible to make checks without funds.

A common approach to solve the problem is to use a central computer which keeps the accounts of each customer. When a customer wants to make a transaction, he communicates with the bank. A secure system can be built using cryptography. (This approach is actually studied by the banks.)

The disadvantages of this kind of approach are the following:

- if the computer is down, no transaction can be made;

- this kind of system requires a communication with the bank for each transaction. This constitutes a very large flow of information at the bank, and it's difficult to apply it for a widespread system (for everyday transactions).

We could try to solve our problem with a network. It would only partly overcome the disadvantages. That's one of the reasons supporting our second constraint. In this paper, we are interested in a solution where no communication with the bank is required.

The use of a directory in a solution would also be inconvenient. The directory would be huge, and it would be impossible to keep it continuously up to date. The consultation of the directory in itself is a problem if we want to use the system for everyday transactions.

ASSUMPTIONS

We first set up the assumptions under which our solutions can be proved to be correct.

1- A and B are customers of a bank.

2- To become a customer, one has to communicate with the bank in a secret way.

3- In the protocol to become a customer, A provides the bank with a unique identifier IdA. (No two customers have the same identifier.)

4- The bank chooses a public-key cryptosystem (PKC) and publishes it (or just makes it available to every customer).

5- The PKC has the following properties, given a pair of public and secret keys PK and SK:

a) $E_{PK}(D_{SK}(M)) = D_{SK}(E_{PK}(M)) = M$ (see notation below)

b) Even knowing PK, it's infeasible for someone not knowing SK to find $D_{SK}(Y)$, for a given message Y.

c) It's possible to put messages in a special form (publicly known) such that it's infeasible to find any pair (C,M) where M is a message in the special form and $D_{SK}(M) = C$. To tell that a message is in the special form, we will put it between "--". (Ex: $D_{SK}(--"hello"--)$)

(This last property is to keep an enemy from creating any signed message.)

Even though no PKC has been proved to have these properties, some PKC are believed suitable [2,3].

6- A proof system is characterized by a verification function V that is easy to calculate, and such that V(P,E) is true implies that the sentence E is false with a reasonably low probability (directly related to the infeasibility in the PKC).

We will call P a proof of E. For our use, $P = D_{SK}(--S--)$ can be considered as a proof of the sentence "if someone not knowing the secret key SK knows P, then someone knowing SK agreed on the sentence S".

7- The bank chooses a pair of public and secret keys PKS and SKS, according to the published PKC. PKS is known by every customer.

8- The branches of the bank are customers of the bank.

9- The bank is reliable.

10- No secret keys are revealed.

The last two restrictions may seem a little unreasonable. They are nevertheless necessary to rigorously prove the correctness of our solutions. We discuss later the effect of weakening these assumptions.

Now, we can state the problem more formally.

---

notation: $E_{PK}(M)$ is the encoding of M according to the public key PK

$D_{SK}(M)$ is the decoding (or signature) of M according to the secret key SK

Problem

Build a transaction system with the following essential properties:

1- A customer A can make a transaction with any other customer B. (At the same time, B makes a transaction with A).

2- The customer A can prove to a third person that he made the transaction with B, and that B is also a customer of the bank.

3- It's infeasible for a customer to create money. (The total amount of money in the system should always be the same, unless the bank introduces new money.)

4- No communication with the bank is required to make a transaction.

5- No directory of customers is available.

The system should also meet as nearly as possible the following desirable properties:

6- It's infeasible for a listening enemy to guess the terms of a transaction.

7- It's infeasible for a listening enemy to prove to another person that a transaction happened, unless it has been proved to him before.

8- It's infeasible to make a transaction without having the proper funds.

9- A transaction can be made over the phone.

10- Minimize the problems that could happen if a transaction doesn't terminate properly.

11- Minimize the problems, if some information or material is lost.

12- Minimize the problems, if some information or material is stolen.

SOLUTIONS

In what follows, we will see two solutions, each having the essential properties. Nevertheless, each of these will also have serious disadvantages (lack of some desirable properties).

Then, we will combine these two solutions in a third solution which has almost all the desirable properties.

After all, we will mention a fourth solution, better than the third one on certain points, but having more serious disadvantages.

First Solution (Basic)

The following steps are included in the protocol that makes A a customer of the bank:

- A finds a pair of public and secret keys PKA and SKA, for the PKC published by the bank.

- A reveals PKA to the bank, with an effective and an expiration date EF and EX, and an identifier IdA which is unique for the bank (no two customers have the same identifier).

- The bank gives to A a public key certificate:

$$D_{SKS}(\text{--IdA, "public key is PKA", EF, EX--}).$$

When A wants to make a transaction with B, they first exchange and verify their public key certificates. Then A and B, using the keys PKA and PKB to exchange messages, use a contract signing protocol to sign the contract: "IdA makes the transaction T with IdB", where T contains at least the amount of the check and a time stamp.

A and B keep each other's public key certificates with their signed contracts for proof of the transaction. B can cash the check when he goes to the bank (showing his proof).

We can prove that the scheme is good, assuming the contract signing protocol is good, with the following short argument. Only property 2 in the problem has to be shown. A and B each have a proof of the contract between IdA and IdB. By assumption, A doesn't know SKS. The proof that the bank accepted IdA and IdB as customers, with the keys PKA and PKB, is the fact that A and B know the messages certifying the keys.

Contract Signing Protocol

A simple way to implement a contract signing protocol is to exchange a preliminary agreement from B. Suppose A and B want to sign a contract C. First, B sends the message $E_{PKA}(D_{SKB}(\text{--Accept, C--}))$ to A. Then, A sends $E_{PKB}(D_{SKA}(\text{--Sign, C--}))$ to B, and B finally sends $E_{PKA}(D_{SKB}(\text{-- Sign, C--}))$ to A. While the first message is not considered as a signed contract, A can always call the court to force B to sign it.

The problem is that after the first message is sent, B is partly committed to the contract in the sense that A can still delay his decision to sign the contract or not. We could fix a delay after which

the preliminary agreement is cancelled. This would be acceptable for
our use, since signing a check is usually a one-way commitment. But
there are much nicer contract signing protocols. Refer to [4,5,6].

The disadvantage of our first solution is that checks without
funds can be made easily. This could be catastrophic if the secret
key of a customer is stolen, since there is no way to keep the thief
from making checks.

Second Solution (Electronic money holder)

The second solution uses microprocessors. Each customer has a
microprocessor which contains a certain amount of money. Ideally,
if A wants to make a transaction with B, they connect their MP (microprocessor) together, directly or through a phone line. The money is
transferred from the memory of one MP to the memory of the other. It
should be infeasible to increase the amount of money in a MP without
decreasing the amount of money in another MP by an equal amount.

In order to prove that the method is correct, we have to make a
new assumption:

- It's possible to hide some information in a MP so that the only
  way to know about this information is by the external pins of
  the MP.

This assumption, again, is a little unreasonable. Later we will
investigate weakening this assumption.

Method. In the protocol that makes A a customer of the bank,
in addition to the steps described in the first solution, the following steps are included:

- The bank finds a pair of public and secret keys MPPKA and MPSKA
  according to the PKC. MPSKA is not revealed, even to A.

- The bank produces and gives to A a MP (containing MPSKA) able
  to make a transaction. In each MP, there is a sequence number
  that is incremented after each transaction.

When A wants to make a transaction with B, the following steps
are observed:

- They exchange and verify their respective public keys as in the
  first solution.

- They form the message $C = D_{SKA}(D_{SKB}(--\text{"IdA makes a transaction}$
  T with IdB, sequence numbers are seqA and seqB"--)), which is
  the contract to be signed by each MP. Until now, nobody is
  committed. There is no trouble if A refuses to sign the

message signed by B.

- Each MP, after exchanging and verifying bank certificates for PKA, PKB, MPPKA and MPPKB, enters in a contract signing protocol to sign the contract C.

- After the contract is signed, A's MP decreases its internal amount of money, increments its sequence number and produces the decreasing certificate message:

    D_{MPSKA}(--"IdA to IdB", seqA, seqB, amount--).

- A sends the latter message, signed with SKA and encoded with MPPKB to B. When the coded message is entered in B's MP, the internal amount of money is increased and the sequence number is incremented.

A proof of the transaction consists of a proof of the contract, the decreasing certificate message signed with SKA, and the bank certificates of the keys.

Comments. It was important to make the increase of money in one MP directly dependent on the decrease of money in the other MP. If it was not the case, in either of the above-mentioned contract signing protocols, A and B could collaborate to increase the money in one MP without decreasing the money in the other.

In this second solution again, we can prove (in the same way) that the requirements of our problem are met. Also, if ever A doesn't want to send the last message, B can go to a court and force A to do so, having a proof of the transaction. The court would then charge to A and B a percentage of the amount of the transaction. This would discourage A from not sending the message and discourage B from going to the court just to give some trouble to A.

We should notice that the bank still doesn't know SKA and is not able to produce a proof or a check instead of a customer.

The main disadvantage of this method is that if we lose a MP, the money is lost. A thief, even if he doesn't know the password, is able to destroy the money. Nevertheless, one doesn't have to keep a lot of money in the MP since it's easy to call a branch of the bank and withdraw some money. Another problem is that if ever someone is able to read inside his MP, he could produce another MP with an infinite amount of money, and the trick would be difficult to trace by the bank.

Third Solution (Combination of first and second)

We can combine the two methods in a third method which overcomes most of the disadvantages mentioned earlier.

The new method is exactly like the second one, except that the official amount of money is kept at the bank as in the first method. One must present the proofs of checks to the bank to really have the money.

When A wants to cash his checks, he brings all the checks corresponding to the sequence numbers of his MP that have not yet been treated by the bank. The bank verifies the sequence numbers on the checks with the actual sequence number of A's MP. The bank then takes the money from the other customers' accounts. The checks written by A are not treated, but only kept in the bank's records until they are cashed (if not already cashed).

There is no restriction on how the proofs of checks are kept until they are presented to the bank. The most practical way is to keep the proofs in a memory of the MP. This memory doesn't need to be in a secure part of the MP. If the memory is not large enough, the proofs could be kept on a piece of paper.

In a practical application, we can almost assume that a check is never lost unless the MP is also lost. However, if ever a check is not cashed, the bank should be able to adjust the amount of money in each MP. If we give six months to someone to cash a check, and if A writes a check that is not cashed after six months, the bank can then adjust A's MP. (The bank knows the amount because A provided a proof that the check was written.)

If a customer doesn't have all the proofs for his sequence numbers, then the bank keeps a record of the date and the difference between the amount of money in the MP and in the account. Whenever someone cashes a check written by A before this date, the record is adjusted. After six months, the difference of money can be given or charged to A, depending on the sign of the difference. For example, suppose A wrote a check for $40, received a check for $30, and lost the proofs. The bank account would be $10 higher than the MP's amount of money (before the $40 check is cashed). After six months, if the $40 check had been cashed, the bank would know that A had lost checks for the total amount of $30, and would decrease A's MP by $30. If the check had not been cashed, the bank would increase A's MP by $10.

If someone loses a check, he can ask for a replacement. For example, if A loses a check written by B, he can ask for his proof of the transaction. If B lost it too, nobody has a proof and the check is lost. B would have to write another check, with a note saying that the bank shouldn't accept the check if the real check is cashed, and vice versa. (hopefully, B is able to clearly identify the lost check.)

The case of a lost MP is a little more delicate. After six months, the bank knows exactly the amount of money in the lost MP. But this is assuming that the MP is really lost. If the bank gives a new MP with the money to a customer who has lost his MP, this customer is able to make checks without funds if he finds his old MP. There are many possible solutions to this problem. In one of these, the MP requires a certified permission from the bank to be able to produce a check. A certificate has the form:

$$D_{SKS}(\text{--"The MP belonging to IdA is allowed to make checks until a specific date"--}).$$

The bank can let the customer decide the frequency with which he has to ask for a new certificate. The new certificate can be sent by mail or by telephone. If a MP is lost, the bank begins to count the six months period after the last day of permission.

Let's now investigate robberies. If a MP is stolen with the password, the thief will be able to use the money inside it without much risk. He will not ask the bank for more money, because the bank will have been informed (hopefully).

The bank's MP can be made almost impossible to steal, since there is no need to have direct access to it. Also, the bank robber can't ask the bank to give him a certain amount of money, because the robber's identity would be revealed. The only way is for the robber to ask for a bank certificate. For this, the key SKS is needed, and this key is not known at any branch. This means that an account can't be opened instantly at a branch. The branch has to send the request to the "certifying center".

We just saw that the security of the system relies almost entirely on the central key SKS. Indeed, someone knowing SKS could make his own MP, his own certificate, and make transactions without ever being suspected. It's then important that none of the branches know SKS and that this key is protected by all possible means.

Another problem with this method is that it is possible to make a check without funds in some circumstances. Indeed, the amount of money in the MP is not always the official amount. If someone loses checks, the money is still available in the MP. He is able to make checks with the funds recorded in his MP but actually lost. This is not really annoying, because usually a lost check can be replaced, and as we mentioned, a check would probably never be lost.

Fourth Solution (Traveler's Checks)

This solution, having many disadvantages, can be useful for some applications, because of its simplicity.

The solution is an adaptation of the actual Traveler's Checks system. The customer registration method is the same as in the previous methods. First, the bank produces the checks, on a paper form, with a nominal value X and a serial number N. Then, the bank prints on the check a message (signed with SKS) indicating the branch it belongs to.

When a customer A wants to give the check to another customer B (A or B could be branches), they first enter in a contract signing protocol to sign the transaction (which of course mentions the serial number N). They then print on the check the proof of the transaction.

Since the method includes our first solution, the essential properties are met (except partially property 4, where we have to go to the bank to get the checks, or cash one when there is no more space on it).

The fact that each check has a nominal value is a disadvantage for our application. But we can use this solution in an application where the transactions always have some fixed values.

Many of the desirable properties are not met. Since there are many proofs on the check, desirable properties 6 and 7 are not met. Since we use a physical support for the check form, property 9 is not met.

In this method, the only way to use money without having the funds is to produce a false check. It's then important for this method to use paper (or some other physical support) for the check. If the checks were only messages, it would be our first solution with additional restrictions. Now, a false check can't be made by accident (it's easier to duplicate a message than a check form).

It's also important that all the proofs remain on the check or else two customers could collaborate to make false checks by making many copies of a check, and never be discovered by the bank, since the proofs would be erased.

CONCLUSION

We have found a way to apply public-key cryptography in a banking system. This system has only a few disadvantages, fewer than the system now in use. But we still have to find a provably good cryptosystem. Would people be willing to build a banking system on a cryptosystem which is only conjectured to be secure?

REFERENCES

[1] Longpré, L., "Etude des applications de la cryptographie à clef publique", Mémoire de maitrise, Université de Montréal, Département d'informatique et de recherche operationnelle, Août 81.
[2] Diffie, W. and Hellman, M.E., "New Directions in Cryptography", IEEE Transactions on Information Theory, Vol. IT-22, Nov. 1976, pp. 644-654.
[3] Rivest, R.L., Shamir, A. and Adleman, L., "On Digital Signatures and Public-Key Cryptosystems", Communications of the ACM, Vol. 21, No. 2, Feb. 1978, pp. 120-125.
[4] Blum, M., "How to Exchange (Secret) Keys", Memorandum No. UCB/ERL M81/90, March 1982.
[5] Even, S., Goldreich, O., and Lempel, A., "A Randomized Protocol for Signing Contracts", Technical Report #233, TECHNION Israel Institute of Technology, Feb 1982.
[6] Rabin, M.O., "Transaction Protection by Beacons", Doc. TR-29-81, November 1981, Aiken Computation Laboratory, Harvard University.

BLIND SIGNATURES FOR UNTRACEABLE PAYMENTS

David Chaum

Department of Computer Science
University of California
Santa Barbara, CA

INTRODUCTION

Automation of the way we pay for goods and services is already underway, as can be seen by the variety and growth of electronic banking services available to consumers. The ultimate structure of the new electronic payments system may have a substantial impact on personal privacy as well as on the nature and extent of criminal use of payments. Ideally a new payments system should address both of these seemingly conflicting sets of concerns.

On the one hand, knowledge by a third party of the payee, amount, and time of payment for every transaction made by an individual can reveal a great deal about the individual's whereabouts, associations and lifestyle. For example, consider payments for such things as transportation, hotels, restaurants, movies, theater, lectures, food, pharmaceuticals, alcohol, books, periodicals, dues, religious and political contributions.

On the other hand, an anonymous payments systems like bank notes and coins suffers from lack of controls and security. For example, consider problems such as lack of proof of payment, theft of payments media, and black payments for bribes, tax evasion, and black markets.

A fundamentally new kind of cryptography is proposed here, which allows an automated payments system with the following properties:

(1) Inability of third parties to determine payee, time or amount of payments made by an individual.

(2) Ability of individuals to provide proof of payment, or to determine the identity of the payee under exceptional circumstances.

(3) Ability to stop use of payments media reported stolen.

BLIND SIGNATURE CRYPTOSYSTEMS

The new kind of cryptography will be introduced first in terms of an analogy and then by description of its parts, their use, and the resulting security properties. No actual example cryptosystem is presented.

Basic Idea

The concept of a blind signature can be illustrated by an example taken from the familiar world of paper documents. The paper analog of a blind signature can be implemented with carbon paper lined envelopes. Writing a signature on the outside of such an envelope leaves a carbon copy of the signature on a slip of paper within the envelope.

Consider the problem faced by a trustee who wishes to hold an election by secret ballot, but the electors are unable to meet to drop their ballots into a single hat. Each elector is very concerned about keeping his or her vote secret from the trustee, and each elector also demands the ability to verify that their vote is counted.

A solution can be obtained by use of the special envelopes. Each elector places a ballot slip with their vote written on it in a carbon lined envelope; places the carbon lined envelope in an outer envelope addressed to the trustee, with their own return address; and mails the nested envelopes to the trustee. When the trustee receives an outer envelope with the return address of an elector on it, the trustee removes the inner carbon lined envelope from the outer envelope; signs the outside of the carbon lined envelope; and sends the carbon lined envelope back, in a new outer envelope, to the return address on the old outer envelope. Thus, only authorized electors receive signed ballot slips. Of course, the trustee uses a special signature which is only valid for the election!

When an elector receives a signed envelope, the elector removes the outer envelope; checks the signature on the carbon lined envelope; removes the signed ballot slip from the carbon lined envelope; and mails the ballot to the trustee on the day of the election in a new outer envelope, without a return address.

When the trustee receives the ballots, they can be put on public display. Anyone can count the displayed ballots and check the signatures on them. If electors remember some identifying aspect of their ballot, such as the fiber pattern of the paper, they can check that their ballot is on display. But since the trustee never actually saw the ballot slips while signing them (and assuming every signature is identical), the trustee can not know any identifying aspect of the ballot slips. Therefore, the trustee can not know anything about the correspondence between the ballot containing envelopes signed and the ballots made public. Thus, the trustee can not determine how anyone voted.

Functions

Blind signature systems might be thought of as including the features of true two key digital signature systems combined in a special way with commutative style public key systems. The following three functions make up the blind signature cryptosystem:

(1) A signing function s' known only to the signer, and the corresponding publically known inverse s, such that $s(s'(x))=x$ and s give no clue about s'.

(2) A commuting function c and its inverse c', both known only to the provider, such that $c'(s'(c(x)))=s'(x)$, and $c(x)$ and s' give no clue about x.

(3) A redundancy checking predicate r, that checks for sufficient redundancy to make search for valid signatures impractical.

Protocol

The way these functions are used is reminiscent of the way the carbon paper lined envelopes were used in the example described above:

(1) Provider chooses x at random such that $r(x)$, forms $c(x)$, and supplies $c(x)$ to signer.

(2) Signer signs $c(x)$ by applying s' and returns the signed matter $s'(c(x))$ to provider.

(3) Provider strips signed matter by application of c', yielding $c'(s'(c(x)))=s'(x)$.

(4) Anyone can check that the stripped matter $s'(x)$ was formed by the signer, by applying the signer's public key s and checking that $r(s(s'(x)))$.

Properties

The following security properties are desired of the blind signature system comprising the above functions and protocols:

(1) Digital signature--anyone can check that a stripped signature $s'(x)$ was formed using signer's private key s'.

(2) Blind signature--signer knows nothing about the correspondence between the elements of the set of stripped signed matter $s'(x_i)$ and the elements of the set of unstripped signed matter $s'(c(x_i))$.

(3) Conservation of signatures--provider can create at most one stripped signature for each thing signed by signer (i.e. even with $s'(c(x_1))$ ... $s'(c(x_n))$ and choice of c, c', and x_i, it is impractical to produce $s'(y)$, such that $r(y)$ and $y \neq x_i$).

As is common in cryptographic work, the possibility that the same random number could be generated independently is ignored.

UNTRACEABLE PAYMENTS SYSTEM

An example payment transaction will illustrate how the blind signature systems introduced above can be used to make an untraceable payments system. The critical concept is that the bank will sign anything with its private key, but anything so signed is worth a fixed amount, say $1. The actors in the example below are a bank, a payer, and a payee. A single note will be formed by the payer, signed by the bank, stripped by the payer, provided to the payee, and cleared by the bank. The following traces the detailed steps of a single payment transaction:

(1) Payer chooses x at random such that $r(x)$, and forms note $c(x)$.

(2) Payer forwards note $c(x)$ to bank.

(3) Bank signs note, i.e. forms $s'(c(x))$, and debits payer's account.

(4) Bank returns the signed note, $s'(c(x))$, to payer.

(5) Payer strips note by forming $c'(s'(c(x)))=s'(x)$.

(6) Payer checks note by checking that $s(s'(x))=x$ and stops if false.

(7) Payer makes payment some time later by providing note $s'(x)$ to payee.

(8) Payee checks note by forming $r(s(s'(x)))$ and stops if false.

(9) Payee forwards note $s'(x)$ to bank.

(10) Bank checks note by forming $r(s(s'(x)))$ and stops if false.

(11) Bank adds note to comprehensive list of cleared notes and stops if note already on list.

(12) Bank credits account of payee.

(13) Bank informs payee of acceptance.

Notice that by the blind signature property above, when the bank receives a note to be cleared from the payee in step (9) the bank does not know which payer the note was originally issued to in step (4). The digital signature and related conservation of signatures properties above ensure that counterfeiting is not possible.

Auditability

Extension of current practice suggests that payers receive digital receipts from payees. These receipts would include the usual description of the goods or services purchased, and the date. In addition, the receipt could also include a copy of the note. Under exceptional circumstances, such as an audit, the note would allow the payer, with the cooperation of the bank (and clearing house(s) as described below), to verify which account the note was actually deposited to.

A receipt indicating that a note was deposited to an account other than the account actually deposited to would be evidence of fraud. One dissatisfied customer of a black market could reveal a note supplied to the black market, which could then be traced to the account it ultimately ended up in. Uncleared notes reported as stolen could be included on clearing house lists and thus be prevented from being cleared; stolen notes cleared could be traced.

Receipts issued by payee to payer provide control over all outflows, and thus all flows of funds. A taxpayer could provide verifiable receipts for any expenditures needed for tax audit. Individuals could be required to keep receipts for substantial inflows, but inflow receipts maintained by organizations may be undesirable, if they could reveal the organization's patrons.

Elaborations

The simple system of the above example could be extended in various ways to provide economy of mechanism, disaggregation of services, and decentralization. For example, obvious efficiencies would result from use of multiple denomination notes. The banking and clearing house functions could be separated. There might be multiple banks; multiple clearing houses could serve different or overlapping banks. Periodic changes of the key(s) used to sign notes might increase security, increase auditability, and reduce uncertainty about the size of the money supply.

SUMMARY AND IMPLICATIONS

A new kind of cryptography, blind signatures, has been introduced. It allows realization of untraceable payments systems which offer improved auditability and control compared to current systems, while at the same time offering increased personal privacy.

A RANDOMIZED PROTOCOL FOR SIGNING CONTRACTS

(Extended Abstract)

S. Even[*], O. Goldreich and A. Lempel

Computer Science Department
Technion - Israel Institute of Technology
Haifa, Israel

1. INTRODUCTION

Suppose two parties A, and B, in a communication network, have negotiated a contract, which they wish to sign. To this end, they need a protocol which has the two following properties:

(1) At the end of an honest execution of the protocol, each party has a signature of the other.

(2) If one party, X, executes the protocol honestly, his counterpoint, Y, cannot obtain X's signature to the contract without yielding his own signature.

It was shown by Even and Yacobi [1] that no such deterministic protocol exists without the participation of a third party. Assuming reliable third parties exist, it is still desirable to have a protocol for signing contracts in which no third party is required. Even [2], proposed a protocol based on the puzzle concept of Merkle [3] using any Public Key Cryptosystem (PKCS) deemed secure. Other protocols, relying on the infeasibility of certain number-theoretic operations, such as factoring of large integers, were suggested by Blum and Rabin [4] and Blum [5].

The notion of Oblivious Transfer (OT) was introduced by Rabin [6], with an implementation based on the integer factoring problem. We propose what we believe to be a more natural definition and present an implementation using any PKCS.

---

[*]Supported in part by the Fund for the Promotion of Research at the Technion

We describe a protocol for signing contracts which uses OT. Its advantage over the protocol proposed by Even [2] is that there is neither reliance nor reference to the value of the contract's context.

2. ASSUMPTIONS

We assume the existence of a secure PKCS [7] and that the cost and time of computation is approximately the same for both parties to the contract.

Let E_x and D_x be the encryption and decryption algorithms, respectively, generated by feeding the word x to the key generating algorithm. We assume that for every x and every ω, E_x and D_x are defined and

$$E_x(D_x(\omega)) = D_x(E_x(\omega)) = \omega.$$

We also assume that every participant A, in the network, randomly chooses a word x_A, from which he generates an encryption-decryption pair (E_{x_A}, D_{x_A}) (hereafter denoted by (E_A, D_A)) and announces his encryption key (E_A). Clearly, A can sign a document M by transmitting $D_A(M)$.

We also use a secure conventional cryptosystem F. Its existence is guaranteed by the (assumed) existence of a secure PKCS; however, one can use any trusted conventional system, e.g. the DES [8]. Denote the encryption and decryption algorithms with key k, by F_k and F_k^{-1} respectively.

3. OBLIVIOUS TRANSFER

An <u>Oblivious Transfer</u> (OT) of a recognizable message M is a protocol by which the sender (hereafter denoted by S) transfers to the receiver (hereafter denoted by R) the message M, so that R can read M with probability one half while S has no way of knowing whether R can actually read M.

Formally OT has to satisfy the following axioms:

(i) R can recognize M [e.g. M is a signature on some known message M', i.e. $M = D_S(M')$].

(ii) If S is honest R gets M with a priori probability one half. For S, the posteriori probability that M was actually read by R remains one half.

(iii) If S tries to cheat, R will detect it with probability at least one half.

An implementation of an OT satisfying these axioms is presented in Section 6.

4. THE CONTRACT SIGNING PROTOCOL

The parties to the protocol will be called A and B.

(1) A generates randomly an ordered set $\{x_i\}_{i=1}^n$ of keys for the conventional system F. He declares that if B is able to present (n-m) signed members of the ordered set $\{M_i\}_{i=1}^n$, then he is committed to the contract C, and signs this declaration. B acts symmetrically generating the keys $\{y_i\}_{i=1}^n$.

(2) A transmits to B the ordered set $\{F_{x_i}(D_A(M_i))\}_{i=1}^n$.
B transmits to A the ordered set $\{F_{y_i}(D_B(M_i))\}_{i=1}^n$.

(3) <u>for</u> i = 1 to n <u>do</u>
<u>begin</u>
    A sends x_i to B via OT
    B sends y_i to A via OT
<u>end</u>

(4) <u>for</u> j = 1 to ℓ <u>do</u> (ℓ is the length of the keys for F)
<u>begin</u>
    A transmits the j-th bit of every x_i to B
    B transmits the j-th bit of every y_i to A
<u>end</u>

<u>Note:</u> The interleaving in step (3) is not essential.

To avoid being cheated the parties should take the following precaustions:

(a) During step (3) each party, while playing the role of R in OT, should use the cheat-detection mechanism of the OT. [Its existence is guaranteed by axiom (iii). Also note that the keys are recognizable using the information transferred in step (2).]

(b) While executing step (4) each party should check whether the bits revealed to him during the alternating substeps match the bits of the keys actually disclosed to him in step (3). [Note that after step (3) is completed, each party knows, on the average, one half of his counterpart's keys; the latter, however, is oblivious as to which of his keys were actually disclosed.]

A party will stop further execution of the protocol as soon as he detects an attempt to cheat.

5. ANALYSIS OF THE PROTOCOL

If both parties follow the protocol honestly to its conclusion then either will have a signature by the other to the contract C, and will know it. In fact, each party will have all n signed M_i's.

Let PR(n,m) denote the probability that X gets at least n-m keys during the execution of step (3) of the protocol.

<u>Theorem</u>: If $n \geq 100$ and $n-m \geq .78 \cdot n$ then $PR(n,m) < 2^{-(m+1)}$.

Thus, the probability that X will have his counterpart's signature to the contract before the execution of step (4) of the protocol, is less than $(\frac{1}{2})^{m+1}$. (If this occurs X might stop the procedure before his counterpart has X's signature.)

If X decides to cheat Y, he has to make sure that Y gets less than (n-m) signed M_i's. To this end, during the execution of the protocol, X must designate at least m+1 M_i's for which Y is not to have X's signature.

Without loss of generality, assume that X = A. A may prevent B from having the i-th signature (i.e. $D_A(M_i)$) by one of the following actions:

(1) Transfer a "fake" $F_{x_i}(D_A(M_i))$ in step (2).
(2) Cheat in execution of the OT of x_i (in step (3)).
(3) Cheat in the disclosure of the bits of x_i (in step (4)).

Clearly, it makes no sense to take more than one of these three possible actions, since one of the first two suffices to make sure tha B will not get $D_A(M_i)$, while a multiple attempt for the same i may increase the chances of being caught.

By axioms (i) and (iii) of OT, actions (1) or (2) will be detected with probability at least one half; while by axiom (ii), the probability of being caught in action (3) is exactly one half. Thus, the probability that any party will succeed in cheating the other, is at most $(\frac{1}{2})^{m+1}$.

The total risk for X in using the protocol amounts to the sum of two probabilities; the probability that Y gets the signature in step (3) and the probability that Y succeeds in cheating X. This risk is bounded from above by $2(\frac{1}{2})^{m+1} = (\frac{1}{2})^m$.

An important feature of our protocol is that with high probability, $(1-2^{-m})$, the feasibility of obtaining a signature by computation is about the same for both parties. This observation is based on the fact that computing the signature becomes feasible only during the execution of step (4) and at this point each party knows that, with very high probability, he has the information required for the computation of his counterpart's signature. This feature is absent from Even's protocol [2], where the information required for the computation of the signature passes from X to Y, before X is able to verify that he can compute the signature of Y.

It should be noted, however, that if X stopes the correspondence during step (4), his advantage over Y is at most one bit per key. If this is considered too big an advantage, one can change step (4) of the protocol so that only a single bit is transferred at a time instead of n bits.

6. AN IMPLEMENTATION OF OBLIVIOUS TRANSFER

The proposed implementation of an oblivious transfer of a message M from S to R proceeds as follows:

(0) S chooses, randomly, two pairs $\{(E_i, D_i)\}_{i=1}^{2}$ of encryption-decryption algorithms for the PKCS. R chooses, randomly, a key K for the conventional cryptosystem F.

(1) S transmits E_1, E_2 to R.

(2) R chooses, randomly, $i \in \{1,2\}$ and transmits $E_i(K)$ to S.

(3) S chooses, randomly, $j \in \{1,2\}$, computes $K' \triangleq D_j(E_i(K))$ and transmits the pair $(F_{K'}(M), j)$ to R.

Remarks:

(1) Assuming that K looks like random noise and that E_1, E_2 have the same range, S cannot know (or guess with probability of success greater than one half) whether K', computed by him, is the K choosen by R.

(2) By the assumption that the PKCS is secure, R cannot find K' when $K' \neq K$. Due to the security of the conventional cryptosystem, R must know K' in order to read M.

(3) R can read M iff $i = j$. Thus, he can detect cheating by S with probability one half.

(4) In the RSA [9] scheme, distinct E_i's do not have the same range; nevertheless, this can be fixed. Other implementations of OT via RSA were suggested by Rabin and Micali.

7. COMMENT ON EXTENSIONS

Using an (n-m,n) threshold - scheme and generalizing the use of the OT we have developed protocols for:

(1) Sending Certified Mail [Mailing Disclosures],

(2) Coin Flipping [Lottery],

with the same exponentially decreasing probability of being cheated.

REFERENCES

[1] Even, S., and Yacobi, Y., Relations Among Public Key Signature Systems, TR#175, Computer Science Dept., Technion, Haifa, Israel, March 1980.

[2] Even, S., A Protocol for Signing Contracts, TR#231, Computer Science Dept., Technion, Haifa, Israel, January 1982.

[3] Merkle, R.C., Secure Communication Over Insecure Channel, Comm. ACM, Vol. 21, April 1978, pp. 294-299.

[4] Blum, M., and Rabin, M.O., How to Send Certified Electronic Mail. In preparation.

[5] Blum, M., How to Exchange (secret) Keys, Memo No. UCB/ERL M81/90, March 1982. To appear in CACM.

[6] Rabin, M.O., Private communication.

[7] Diffie, W., and Hellman, M.E., New Directions in Cryptography, IEEE Trans. Inform. Theory, Vol. IT-22, No.6, November 1976, pp. 644-654.

[8] Data Encryption Standard, National Bureau of Standards, Federal Information Processing Standards, Publ. 46, 1977.

[9] Rivest, R., Shamir, A., and Adleman, L., A Method for Obtaining Digital Signatures and Public Key Cryptosystems. Comm. ACM, Vol 21, February 1978, pp. 120-126.

ON SIGNATURES AND AUTHENTICATION

S. Goldwasser, S. Micali, and A. Yao

Computer Science Division
University of California
Berkeley, California 94720

1. Introduction

The design of cryptographic protocols using trapdoor and one-way functions has received considerable attention in the past few years [1-8]. More recently, attention has been paid to provide rigorous correctness proofs based on simple mathematical assumptions, for example, in coin flipping (Blum [1]), mental poker (Goldwasser and Micali [4]). It is perhaps reasonable to speculate at this time that all cryptographic protocols can eventually be designed to be provably secure under simple assumptions, such as factoring large numbers or inverting RSA functions are computationally intractable in the appropriate sense.

In this paper, we will study in the above light the basic problem of *digital signature*, which was one of the original applications for trapdoor functions in Diffie and Hellman [3]. It is also important as the ability to sign messages is often a prerequisite in other cryptographic protocols.

Diffie and Hellman [3] proposed the following solution to the signature problem. Let M be the *message space*. Every user A in the public network has an encryption algorithm E_A (as specified by a *public key* K_A) listed in a secure public file, and keeps secret a decryption algorithm D_A (specified by a *private key* K'_A), where $E_A(D_A(m)) = m$ for all $m \in M$. The function E_A is a *trapdoor function*, in the sense that it is computationally hard to compute $D_A(m)$ without knowing the secret key K'_A. For user A to send a signed message of $m \in M$ to another user B, A will compute $D_A(m)$, and send both m and $D_A(m)$ to B. Since nobody else is likely to be able to produce $D_A(m)$, B can compute $E_A(D_A(m)) = m$ to verify that he has received a properly signed message from A and retains $D_A(m)$ as a proof that A has signed m.

---

[*]The first two authors were supported by NSF grant MCS 82-04506. Micali is now with University of Toronto; the present address of Yao is Stanford University.

Several implementations of the above approach have been suggested in the literature [6][8]. For example, in the Rabin signature scheme[6], user A enters in the public file a large composite integer N, which is the product of two large primes, while keeping secret the prime factors as the private key. Let $E_A(m) = m^2 \bmod N$. To sign a message m, A computes $\sqrt{m} \bmod N$ which is easy to do knowing the prime factors of N. On the other hand, the computing of $\sqrt{m} \bmod N$ without knowing the prime factors is very difficult, and has been shown by Rabin [6] to be equivalent to factor N.

Although the above general approach is intuitively attractive, it is not easy to give a precise theoretical foundation for their security. In fact, there are several potential difficulties. Firstly, it cannot be used for a very dense message space; for instance, anyone can easily produce in the "square root" scheme a valid signed message $(m^2 \bmod N, m)$, simply by trying different m's until an $m^2 \bmod N \in M$ is found. Secondly, if the message space is very sparse, it might happen that the function $D_A(m)$ will be easy to compute for $m \in M$; for example, it is conceivable that taking the square root modulo N of m of a special form is easy, even though the general problem is as hard as factoring integers. Finally, there is also the possibilty of forging new signed messages may become easy if an adversary has available some signed messages.

There are modifications that can alleviate some of the above problems. For example, by signing m with $D_A(mu_A)$, where u_A is a special string, it lessens the dense-message-space problem. Still, there has not been any published signature scheme based on the public-key idea that has been proved to be secure under simple mathematical assumptions such as that the factoring of integers. In this paper we will demonstrate this possibility by presenting a signature scheme whose security can be proved under some reasonable assumptions similar to those used in the RSA encryption cryptosystem [8].

We remark that there are other types of signature schemes in the literature, such as Rabin's scheme [7], and Lamport-Diffie scheme as given in [3]. This latter scheme is of special interest, as our signature scheme in Section 4 can be viewed as an improvement of their scheme.

2. Signature Schemes

As our purpose here is to present the new signature method, and to prove the security of this particular scheme, we will only informally describe what a *signature scheme* is. A signature scheme has a parameter n, which can be thought of roughly as the number of bits in the argument of the trapdoor functions employed. We are interested in constructing schemes which become computationally impossible to break when n becomes large. From now on, an efficient procedure will mean a probabilistic polynomial-time algortihm.

Let b be a fixed integer. Assume that each user will sign at most n^b bits of messages. A *signature scheme* S specifies, given n, how a user A generates efficiently a pair of binary strings (P_A, S_A), where P_A is the *public information* to be listed in a public file and S_A is the *private information* to be kept secret. At any time A is capable of producing efficiently from n and S_A a *signed message* m' for any message $m \in M$. It should be easy for any one to compute m from m' and P_A, and verify that the latter is a legitimate signed message of m.

The security requirement we wish to impose is that an adversary cannot hope to forge a new signature, even after seeing a large number of genuine signed messages from A. Let $T = m_1, m_2, \ldots, m_h$ be a sequence of messages from M with at most n^b bits, and let $T' = m'_1, m'_2, \ldots, m'_h$ denote a sequence of corresponding signed messages from A. A *forger* $\mathcal{B}$ is an efficient algorithm which takes P_A, n and a sequence of signed messages $m'_1, m'_2, \ldots, m'_h$ as inputs, and outputs a string y. We say that A is *secure* if, for every forger $\mathcal{B}$, and all T, the probability of producing a valid signed message $y = m'$ with $m \neq m_i$ for all i is $o(1)$ as n goes to infinity.

3. Background on RSA

Let $N = q_1 q_2$ be a product of two large primes, and $s > 1$ be an integer such that $\gcd(s, \phi(N)) = 1$; $\phi(N)$ is the *Euler totient function*, and is equal to $(q_1 - 1) \cdot (q_2 - 1)$. The RSA encryption method [8] is based on the fact that it is easy to compute the s-th root $x^{1/s} \bmod N$ if one knows the factors q_1 and q_2, but appears intractable if the factors are unknown. For the RSA encryption to be effective, two types of assumptions are needed. Firstly, one needs to be able to generate efficiently the public-private key pairs. Secondly, the computing of the s-th root must be hard. We will now state a version of such assumptions.

The following method of generating a composite number N was suggested in [8]. Generate a random n-bit prime a_1; find the first prime in the arithmetic progression $1 + 2a_1, 1 + 4a_1, 1 + 6a_1, \ldots$, and call it $q_1 = 1 + 2d_1 a_1$. Repeat the above process to obtain a second random prime $q_2 = 1 + 2d_2 a_2$. Let $N = q_1 q_2$. Let us call the above the *standard procedure* for generating a composite N with parameter n. The next assumption will guarantee that the procedure terminates in polynomial time. For definiteness, we assume that one seldom needs to search more than n^3 numbers in finding q_1 and q_2.

Assumption 1: With probability $1 - o(1)$, $d_1, d_2 \leq n^3$ as n goes to infinity.

Let $Z_N^* = \{y | 1 \leq y < N, \gcd(y, N) = 1\}$. Let $\mathcal{A}$ be a boolean circuit that takes as inputs three n^3-bit integers N, s, x, and outputs an n^3-bit integer y. An integer N is said to be *unsafe against* $\mathcal{A}$ if there exists a prime $n^3 < s \leq 4n^b + n^3$ such that for input N, s and every $x \in Z_N^*$, the output y is equal to $x^{1/s} \bmod N$. We will say that $\mathcal{A}$ is an ϵ-*root-taker* if a random composite N generated by the standard procedure has at least a probability ϵ to be unsafe against $\mathcal{A}$. Let $C_\epsilon(n)$ be the size of a smallest ϵ-*root-taker*.

Assumption 2: For any fixed $\epsilon > 0$ and any fixed polynomial $q(n)$, $C_\epsilon(n) > q(n)$ as n goes to infinity.

We remark that these assumptions are neither necessary nor sufficient to guarantee the security of the standard RSA encryption [8]; one can, however, easily construct variants of RSA encryption that are secure under these assumptions.

4. A New Signature Scheme Based on RSA

We will define a signature scheme $\mathcal{R}$. Let n be the parameter of the signature scheme. Any user A will have public information $P_A = (N, x)$, where $N = q_1 q_2$ is a random composite number generated by employing the standard procedure with parameter n, and x is a random element uniformly chosen from Z_N^*; the private information is $S_A = (q_1, q_2, x)$.

We now describe how A will sign a sequence of messages $m_1, m_2, \ldots, m_h$. We first define a few terms. Let $p_{t,i}$ denote the i-th smallest prime greater than t. Define

$$f_{N,t}(x, i, 0) = x^{1/p_{t,3i+1}} \bmod N,$$

$$f_{N,t}(x, i, 1) = x^{1/p_{t,3i+2}} \bmod N,$$

$$f_{N,t}(x, i, \#) = x^{1/p_{t,3i+3}} \bmod N.$$

For any binary string $v = b_1 b_2 \cdots b_u$, let $\sigma_{N,x,t}(v) = (t, f_{N,t}(x, 0, \#), f_{N,t}(x, 1, b_1), f_{N,t}(x, 2, b_2), \cdots, f_{N,t}(x, u, b_u), f_{N,t}(x, u+1, \#))$.

To sign the first message m_1, compute the string $\sigma_{N,x,t}(m_1)$, which is to be the signed message m_1'; keep the value of $t_1 = p_{n^3, 3(|m_1|+2)}$ in the memory. Now suppose $m_1, m_2, \cdots, m_k$ have been signed, and the largest prime that has been used is $t_k = p_{n^3, w}$, where $w = 3\sum_{i=1}^{k}(|m_i| + 2)$. The signed message for m_{k+1} will be $\sigma_{N,x,t_k}(m_{k+1})$.

5. Proof of Security

Theorem 1. The signature scheme $\mathcal{R}$ is secure.

We need a result of A. Shamir [9], which we will state informally. Let $s_1, s_2, \ldots, s_w$ be a set of distinct primes, and L be a circuit that takes inputs $N, x, x^{1/s_1} \bmod N, x^{1/s_2} \bmod N, \ldots, x^{1/s_w} \bmod N$ and outputs $y = x^{1/s} \bmod N$ for an ϵ-fraction of x. Then there is a circuit not much larger than L such that it takes inputs N, x and outputs $y = x^{1/s} \bmod N$ for an ϵ-fraction of x.

We will now give a sketch of proof for Theorem 1. Assume the theorem is false. Let $\mathcal{B}$ be a forger such that the probability for producing a new signed message is not $o(1)$ when the signature scheme $\mathcal{R}$ is used. Then, for infinitely many values of n and some fixed $\epsilon > 0$, the following is true. There exists a sequence of messages $m_1, m_2, \ldots, m_h$ of total length at most n^b bits, such that $\mathcal{B}$ will output correctly with probability $\geq \epsilon$ the value of a new root $x^{1/s} \bmod N$ for some prime $s \in S_n$, where S_n is the set of all primes between n^3 and $n^3 + 4n^b$. Let us call N a *favorable* number for $\mathcal{B}$ if the probability of $\mathcal{B}$ outputing a correct root $x^{1/s} \bmod N$ is no less than $\epsilon/2$ for a random $x \in Z_N^*$. A simple probability conservation argument shows that the probability of a random N generated by the standard procedure is favorable for $\mathcal{B}$ must be be at least $\epsilon/2$. For each favorable N let $r(N, n)$ be a prime for which $\mathcal{B}$ outputs $x^{1/r(N,n)} \bmod N$ at least for a fraction of $\epsilon/(2 \cdot 4n^b)$ of the values of $x \in Z_N^*$.

For each prime $s \in S_n$, let V_s be the set of N for which $r(N, n) = s$. It follows that there exists a polynomial-size boolean circuit $E_{s,n}$ with inputs $N, x, x^{1/\alpha} \bmod N, \alpha \neq s$ and an output y such that, for each $N \in V_s$, the output y has a probability at least

$\epsilon/(2 \cdot 4n^b)$ to be $y = x^{1/s} \bmod N$ for a random $x \in Z_N^*$. It can be shown, using Shamir's result mentioned earlier, there exists a polynomial-size circuit $E'_{s,n}$ such that, given inputs N, x with $N \in V_s$, the output is $y = x^{1/s} \bmod N$ for all x. It is straightforward to synthesize all these circuits $E'_{s,n}$ to obtain a polynomial-size circuit D_n with inputs N, x, s and output y satisfying the following condition: For each N that is favorable for $\mathcal{B}$, there is a value of s such that $y = x^{1/s} \bmod N$ for all x. But this means D_n is a polynomial-size $(\epsilon/2)$-*root-taker*, contradicting Assumption 2 in Section 3. This proves the theorem.

Remarks. We have omitted mathematical details in several places. A more comprehensive version, including another new signature scheme, will appear in a forthcoming paper.

References

[1] M. Blum, "Coin flipping by telephone," *Proc. of IEEE, Spring CompCon* 1982, 133-137.

[2] R. DeMillo, N. Lynch, and M. Merritt, "Cryptographic protocols," *Proc. 14th Ann. ACM Symp. on Th. of Comp.*, San Francisco, California, May 1982, 383-400.

[3] W. Diffie and M. E. Hellman, "New directions in cryptography" *IEEE Trans. on Inform. Th.* 22 (1976), 644-654.

[4] S. Goldwasser and S. Micali, "Probabilistic encryption and how to play mental poker keeping secret all partial information," *Proc. 14th Ann. ACM Symp. on Theory of Computing*, May 1982, San Francisco, California, 365-377.

[5] S. Goldwasser, S. Micali, and P. Tong, "How to dstablish a private code on a public network," *Proc. 23rd Ann. IEEE Symp. on Found. of Comp. Sci.*, Oct. 1982, Chicago, Illinois.

[6] M. Rabin, "Digitalized signatures and public-key functions as intractable as factorization," it MIT/LCS/TR-212, MIT Technical Memo, 1979.

[7] M. Rabin, "Digitalized signatures," in *Foundations of Secure Computations*, edited by R. DeMillo, D. Dobkin, A. Jones, and R. Lipton, Academic Press, 1978, 155-168.

[8] R. Rivest, A. Shamir, and L. Adleman, "A method for obtaining digital signatures and public key cryptosystems," *Comm. ACM* 21 (1978), 120-126.

[9] A. Shamir, "On the generation of cryptographically strong pseudo-random sequences," *ICALP* 1981.

Session IV: Applications

CRYPTOGRAPHIC PROTECTION OF

PERSONAL DATA CARDS

Christian Mueller-Schloer
Neal R. Wagner

Siemens AG, Muenchen, Germany
Drexel University, Philadelphia, USA

ABSTRACT

Plastic cards for different types of stored data are in wide use at present. Examples are credit cards and cards bearing access control information for automatic teller machines. More powerful devices with non-volatile read/write memory of several kilobytes, possibly with some intelligence, (Personal Data Cards), open new fields of applications in banking, administration, health care and communications.

If sensitive data is stored on such cards, protection of this data and authentication of the authorized user becomes crucial. This paper describes a method for user verification and selective record protection in a network of terminals and one or more trusted Authentication Servers. The method is based on Single Key and/or Public Key Cryptography in conjunction with personal feature recognition (such as fingerprints) and selective key distribution. All the system information that needs secrecy protection is one key in the Authentication Server(s). The reference pattern for the feature recognition is stored on the card in encrypted form. The Authentication Server(s) can be kept very simple and inexpensive since no long-term data storage is required. As no user specific information remains permanently in the terminals, full user mobility is assured.

INTRODUCTION

Over the past few years plastic cards of the credit card type have been equipped with memory and even some computational capabilities. Besides magnetic storage media, semiconductor memory like ROM or fusable link PROM has been used. Single chip computers have been integrated into cards for handling and safeguarding the stored information [1] (Smart Card, Intelligent Secure Card [2]).

This paper describes a protection mechanism for a hypothetical card (or module) with a large non-volatile read/write memory. A typical implementation could use for example a 1 Mbit bubble memory chip. There is a wide variety of applications of such a Personal Data Card (PDC) [3]. Examples are:

- Electronic ID-card, passport
- Health card, medical history, allergies etc.
- Vehicle log-book
- Debit card
- Access control card
- Cryptographic key for secure communication.

For all these applications a reliable and simple protection system is necessary. The cleartext storage of sensitive data with Personal Identification Number (PIN) checking is potentially dangerous since the checking device can be shortcut and the memory then accessed directly. Storage in encrypted form, on the other hand, requires a verification process which can be carried out by the authorized user only. Directly accessed data is useless without the proper key information. In the following we describe a cryptographic encryption scheme which uses a hybrid system of DES (Data Encryption Standard [4]) and Two Key Cryptography [5,6].

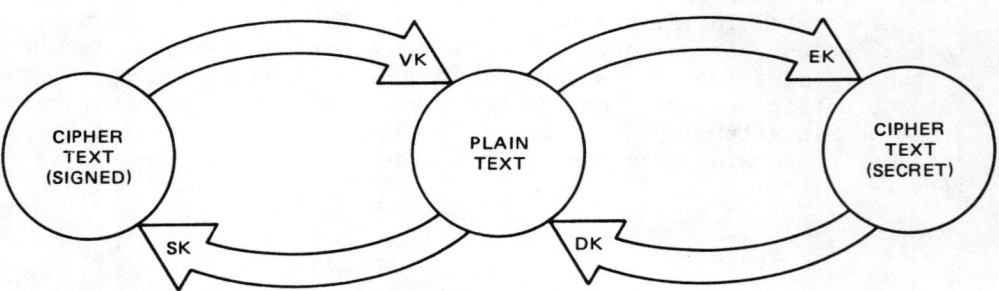

Figure 1. Encryption, Decryption, Signature and Verification in a Two Key Cryptographic System.

Although any Two Key System can be used, we assume the RSA algorithm [7] is used, mainly because of its symmetry between secrecy and signature encryption. In Figure 1 the principle of a Two Key System is given in order to explain our terminology: a plaintext P is encrypted to a ciphertext C with the Encryption Key, EK and decrypted with the Decryption Key, DK. A plaintext can be signed with the Signature Key, SK. The signature is then checked with the Verification Key, VK. The symmetry of RSA resides in the fact that

$$EK = VK \equiv EVK \quad \text{and} \quad DK = SK \equiv DSK.$$

In some systems the Encryption Verification Key, EVK is made public, the Decryption Signature Key, DSK is kept secret (Public Key Cryptosystems). This helps to overcome some of the key distribution and signature problems. We use, in addition, certain system features which are obtained if the distribution of both EVK and DSK is restricted.

Since for the time being fast and inexpensive RSA hardware [8] is not available the DES is used whenever possible to save processing time. Two Key Cryptography is employed wherever a separation of en- and decryption is desirable. An entity M encrypted with a RSA key EVK_i is denoted with brackets:

$$\{M\}EVK_i,$$

and M encrypted with a DES key K_i is written as

$$<M>K_i.$$

SENARIO AND PROTECTION MODEL

The information on the PDC is contained in records. Each record has one specific access right associated with it with respect to any one station in a network. These access rights are:

-     no access
- R   read only
- W   write only
- R/W   read and write

Now we have to find a way to set up and enforce an access policy which controls exactly what every station's access rights to any record are.

One solution would be an access table which has to be stored on each PDC. It has a length of

$$L = r * N_s * 2 \text{ bits}$$

in a network of Ns stations and r records per PDC. For a moderate network of Ns = 5000 stations and r = 20 records per PDC, L = 200 Kbit! This is clearly too much and, in addition, the table approach presents problems as far as updating of the access information is concerned. Besides, the mechanism can be shortcut, since data are in cleartext format.

Before the cryptographic solution is explained let us first define the <u>protection environment</u>: The network consists of stations which can read PDCs and process the information of certain records. The stations are interconnected by a communication system which is assumed non-secure. The stations are grouped into protection <u>domains</u> with the following definition of a domain:

> A domain is the area of responsibility of a certain organization or authority. One and only one domain is the owner of each record. Each domain must have at least one station with write access to its own records. Domains can and generally will overlap each other, i.e. a station can have different access rights to the records of different domains.

For a station being in a domain means to possess a key for the domain's records. There are, however, different types of keys giving different access rights as will be seen in Figure 2.

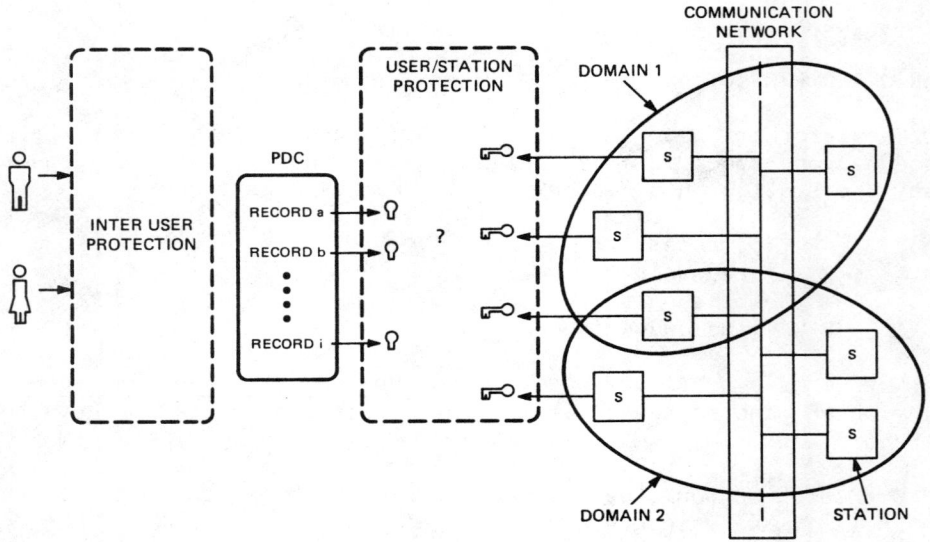

Figure 2. The Protection Model.

Now the <u>protection objectives</u> can be formulated. We want to provide:

1) Protection for the card user against unauthorized stations (<u>User/Station Protection</u>).

   a) Selective protection for records between the domains.

   b) Selective protection for records against stations of the same domain with lower access rights.

2) Protection for the card user against other (unauthorized) card users (<u>Inter User Protection</u>).

Figure 2 illustrates this model.

To classify the access rights of stations and records the following definitions are introduced:

Any station is in one of 3 <u>Station Access Classes</u> (SAC) with respect to every domain. A SAC gives specific access rights to a station. It corresponds to a (mechanical) <u>key</u>.

Any record is in one of 3 <u>Record Access Classes</u> (RAC) assigned by its owner domain. The RAC controls by which station (or group of stations) a record can be accessed. It corresponds to a (mechanical) <u>lock</u>.

Different access rights are assigned to the 9 possible combinations of keys and locks (SACs and RACs) according to Table 1.

Table 1. Access rights associated with the 9 possible combinations of Station Access Classes (SACs) and Record Access Classes (RACs).

	RAC0	RAC1	RAC2
SAC0	R/W	–	–
SAC1	R/W	R	W
SAC2	R/W	R/W	R/W

REALIZATION OF THE PROTECTION POLICY BY KEY DISTRIBUTION AND ENCRYPTION

Every domain (#i) generates one pair of keys of a Two Key System:

EVK_i and DSK_i.

Unlike in a Public Key System none of these is made public. They are both <u>distributed</u> <u>selectively</u> to the stations according to their intended SAC. The records on the PDC are encrypted by their respective owner domains according to their RAC (see Table 2).

Table 3 illustrates how this key distribution/encryption scheme enforces the desired protection policy: A record with RAC1 encrypted under DSK_i can be read by any station having EVK_i (i.e. stations of SAC1 and SAC2) and can be modified only by a station which has DSK_i (i.e. SAC2). Records with RAC2 encrypted under EVK_i can be written by any owner of EVK_i (i.e. SAC1). Stations with SAC0 (usually the majority) can read and modify only unprotected records.

Note that the proposed scheme realizes only a subset of all $4**(N_s*N_r)$ possible protection states since we chose not to use the 4th possible Station Access Class (SAC3) with DSK_i alone distributed to a station. In this case two stations with lower access rights (SAC1 and SAC3) could cooperate to gain SAC2 which is not desirable.

Table 2. Key distribution and encryption according to SACs and RACs.

SAC	station gets		RAC	encryption mode
0	no key		0	no encryption
1	EVK_i		1	encrypted with DSK_i
2	EVK_i, DSK_i		2	encrypted with EVK_i

Table 3. Resulting protection policy.

		RAC0 record	RAC1 {record}DSK_i	RAC2 {record}EVK_i
SAC0	no key	R/W	–	–
SAC1	EVK_i	R/W	R	W
SAC2	EVK_i, DSK_i	R/W	R/W	R/W

EXAMPLE

For illustration purposes let us assume a simplified system with 4 authorities or domains (see Figure 3). Domains 1 and 2 are hospitals with their administrative stations Accounting 1 and 2, respectively. The Insurance Carrier's domain (3) grants certain rights to these accounting stations. Finally, there is the medical domain (4) which includes doctors only. The PDC contains 5 records. Record#1 is unprotected. Record#2 holds insurance information for read access by the accounting stations and read/write access by the Insurance Carrier. Record#3 contains billing information written by the hospitals and to be read/modified by the Insurance Carrier. The data in Record#4 are exclusively for internal use of Hospital 1, and Record#5 holds a health profile for read/modify use by doctors. The records are encrypted and keys distributed according to Figure 3. Table 4 summarizes the realized protection scheme.

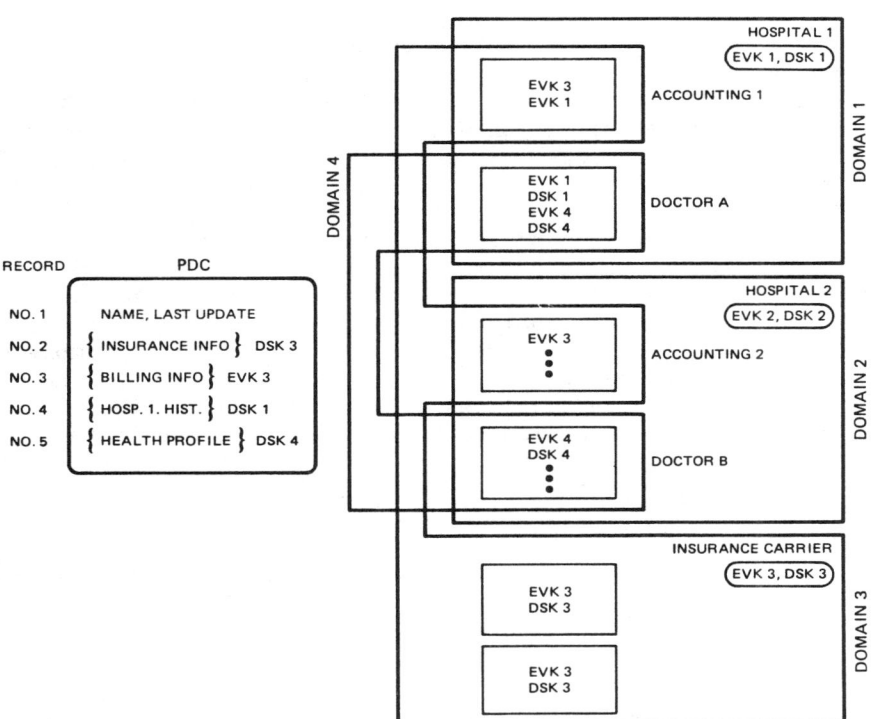

Figure 3. Example: A system with 3 hospitals and one insurance carrier.

Table 4. Resulting protection for the hospital example.

station > record#	Acct. 1	Doctor A	Acct. 2	Doctor B	Insurance
1	R/W	R/W	R/W	R/W	R/W
2	R	–	R	–	R/W
3	W	–	W	–	R/W
4	R	R/W	–	–	–
5	–	R/W	–	R/W	–

PROTECTION AGAINST UNAUTHORIZED CARD USERS

The second part of the protection mechanism provides a way to prevent misuse of lost or stolen cards. For this any personal feature of the authorized user that cannot be copied is sufficient. Voice samples, fingerprints, hand geometry or signature dynamics are examples. Memorized passwords are suited as well if they are long enough [9,10]. For the authentication procedure the personal feature supplied to a station together with the PDC must be compared to a reference feature stored somewhere in secret. The stations themselves cannot hold the reference features of every single user. In addition this would make each station a sensitive area. On the other hand, it is not desirable to store reference features in a central location in order to keep central facilities as simple (and hence as secure) as possible. It is also desirable to have no secret information (like keys) residing permanently in the stations: a station should become dedicated to a specific user only during the interaction period. Afterwards all user specific and secret information must be deleted.

These goals can be achieved by (1) storing reference features in encrypted form right on the PDC and (2) sending feature and encrypted reference feature to a trusted Authentication Server (AS) for verification.

The AS first generates a pair of RSA-keys, the public network key PK.N and the secret network key SK.N [11]. PK.N is then distributed to the domains and their stations. All that is required here is that PK.N cannot be modified. SK.N is kept secret in the AS. The PDC contains information according to Figure 4.

```
                RECORD
                NO. 1     USER NAME, ...
                NO. 2     <PROTECTED RECORDS>K.R
                NO. 3     <REFERENCE FEATURE, K.R >K.N
```

Figure 4. The contents of the PDC.

The selectively protected records are encrypted with the DES-key K.R, and K.R along with the reference feature is encrypted with a secret DES-key K.N kept also by the AS.

When a user supplies his personal feature along with the PDC to a station the feature is encrypted under PK.N together with the encrypted reference feature (#3 in Figure 4) and a time stamp [12] and is sent to AS. The AS decrypts twice (with SK.N and K.N), compares feature and reference feature and returns the record key K.R encrypted and time stamped to the station if all checks have been positive. There K.R is used to decrypt the records (#2 in Figure 4). The station is responsible for deleting all sensitive information (e.g. K.R) after the user interaction has been completed. For details see the card validation protocol below.

CARD VALIDATION PROTOCOL

```
    Station (S)                          Authentication Server  (AS)

read PDC

extract personal feature(s)

generate a temporary key
  K.TEMP at random

generate time stamp
  information TS

compose message M1:
  M1 := (feature,
  <ref.feat.,K.R>K.N,
  TS, K.TEMP)

encrypt M1 under PK.N:
  C1 := {M1}PK.N

send C1 to AS           ----> decrypt with SK.N
```

```
                              check validity of TS

                              decrypt <ref.feat.,K.R>K.N

                              compare feature with ref.feat.
                                                      no
                                  checks postive?  ----> STOP


                              generate TS' from TS using a
                                publicly known function

                              compose message M2:=(TS',K.R)
                                and encrypt with K.TEMP:
                                C2 := <M2>K.TEMP

decrypt with K.TEMP          <----  send C2 to S

check validity of TS'
                no
    check positive? ----> STOP


decrypt <protected records>K.R

    ...
    ...
    ...

delete all transaction specific
   information
```

SUMMARY

It has been shown how Personal Data Cards (PDCs) can be protected effectively by cryptographic mechanisms so that PDCs can be used only by an authorized person, and moreover, the data contents remain secret even if physical access should be gained by an intruder. The card validation requires predistribution of only one public key, and the central authority can be kept very simple since no long term storage capabilities (e.g. for user passwords) are necessary. From the viewpoint of user verification all stations are equivalent so that full user mobility is guaranteed. The different records on the PDCs are protected selectively. The assignment of individual access rights by selective key distribution allows one to enforce complex access policies.

REFERENCES

[1] The Nilson Report, Issue 257, April 1981.
[2] Meyer, C.H., Matyas, S.M., "Some Cryptographic Principles of Authentication in Electronic Funds Transfer Systems", Proceedings of the Seventh Data Communications Symposium, ACM and IEEE, 1981, pp. 73-88.
[3] "The Memory Card -- Applications, Markets, Opportunities", Battelle Study, August 1981.
[4] "Data Encryption Standard", National Bureau of Standards, Federal Information Processing Standard (FIPS) Publication No. 46, Jan. 1977.
[5] Lagger, H., Mueller-Schloer, C., Unterberger, H., "Security Aspects of Computer Controlled Communication Systems", (in German), Elektronische Rechenanlagen, 22 (1980), 6, pp. 276-280.
[6] Hellman, M.E., "The Mathematics of Public Key Cryptography", Scientific American, Vol. 241, No. 2, August 1979.
[7] Rivest, R.A., Shamir, A., Adleman, L., "A Method for Obtaining Digital Signatures and Public Key Cryptosystems", Communications of the ACM, 21 (1978), 2, pp. 120-126.
[8] Rivest, R.A., "A Description of a Single-Chip Implementation of the RSA Cipher", Lambda, 1 (1980), 3, pp. 14-18.
[9] Mueller-Schloer, C., Wagner, N.R., "The Implementation of a Cryptography-Based Secure Office System", Proceeding of the 1982 National Computer Conference, Houston, Texas, pp. 487-492.
[10] Wagner, N.R., "Practical Approaches to Secure Computer Systems", Technical Report UH-CS-81-3, Computer Science Department, University of Houston, Texas, April 1981.
[11] Needham, R.M., Schroeder, M.D., "Using Encryption for Authentication in Large Networks of Computers", Communications of the ACM, 21 (1978), 12, pp. 993-999.
[12] Denning, D.E., Sacco, G.M., "Time Stamps in Key Distribution Protocols", Communications of the ACM, 24 (1981), 8, pp. 533-536.

NON-PUBLIC KEY DISTRIBUTION

Rolf Blom

Linköping University
Dept. of Electrical Engineering
S-581 83 Linköping
Sweden

INTRODUCTION

Assume that it should be possible to protect messages transmitted in a N-user network by encryption. The encryption can either be performed by a public-key crypto system or by a conventional cipher. In the first case there is no need for key distribution. In the second case we have two choices, either to distribute keys from a key distribution center or use a public key distribution algorithm.

But sometimes a point is made of the fact that even the most generally accepted public-key systems (the RSA system, the knapsack and α^x) in principle are invertible when the open key is known. Thus it is interesting to investigate if other schemes exist which have greater theoretical security and at the same time have small demands on storage space. Our main concern is to construct schemes that give a unique key for each user pair.

We will assume that there exists a key distribution center and that user keys or secret data used to generate keys are sent from the center to the users in a secure way.

PROBLEM STATEMENT

The network has N users, and every message transmitted in the network should be enciphered with a key of M bits, unique for the source destination pair. We want to construct a key scheme that requires storage of the least possible number of bits at each user. The number of bits required will be called the size of the <u>user</u>

storage and it will be denoted S. Naturally the key scheme should be such that a user only can derive keys for his own legal use.

REFERENCE SCHEME

The most straight-forward way to perform the key distribution is to give each user N-1 different keys, one for each possible destination in the network. We assume that the same key is used for transmission both ways between two users, then S_R = (N-1)*M bits. From a security point of view this system is the strongest possible since each key can be chosen independently but it also has the largest requirement on user storage.

SUBGROUP SCHEMES

Our original idea was to generate the N-1 keys for each user as a linear combination of $^2\log N$ base elements. Then the user storage should only require $O(\log N)$ bits. And indeed we have found examples of such schemes for N=4, N=8. Fig. 1 shows how all N(N-1)/2 keys can be generated for the case N=8. There we see that e.g. all the keys to be used by source 4 can be generated as linear combinations of (a+b), (a+c) and e (all elements, lie in $GF(2^M)$).

Keys		Destination						
SOURCE		1	2	3	4	5	6	7	8
	1	—	a	b	a+b	f	a+b+f	a+f	b+f
	2	a	—	c	a+c	d	a+d	a+c+d	c+d
	3	b	c	—	b+c	d+f	b+d+f	c+d+f	b+c+d+f
	4	a+b	a+c	b+c	—	e	a+b+e	a+c+e	b+c+e
	5	f	d	d+f	e	—	e+f	e+d	d+e+f
	6	a+b+f	a+d	b+d+f	a+b+e	e+f	—	a+d+e+f	b+d+e
	7	a+f	a+c+d	c+d+f	a+c+e	e+d	a+d+e+f	—	c+e+f
	8	b+f	c+d	b+c+d+f	b+c+e	d+e+f	b+d+e	c+e+f	—

Fig. 1. Keys for a network with 8 users. a,b,c,d,e,f $\in GF(2^M)$.

Observe that if user 1 and 2 cooperate, they can derive all keys for user 3. The keys of source 3 to destination 4 though 8 are the sum of the corresponding keys of source 1 and 2. Hence we have to evaluate the security of systems like this in terms of e.g. how many legal users have to cooperate to be able to produce an extra key, i.e. a key which they don't have legal access to.

It is also necessary to analyse if it is possible to number all users in such a way that it is possible to give a simple algorithm for specifying the linear combination of bais elements for a given destination. It is not obvious how to do this in the example.

Due to the limitations shown by the example and our belief that it would be hard, if possible at all, to generate subgroup schemes with acceptable security, we have not pursued this idea.

A SCHEME BASED ON MDS CODES

To circumvent the difficulties of subgroup schemes we have designed a more practical one. Let us first illustrate the idea in its most simple form. Assume that there are $N=b^\ell$ users in the network and that each user is defined by a unique user number in the range 0 to N-1. The address of user i can then be expressed as a vector $\underline{a}_i = (a_{i0}, a_{i1}, \ldots, a_{i(\ell-1)})$ containing the user number in radix b, i.e.

$$i = \sum_{m=0}^{\ell-1} a_{im} b^m$$

We also define ℓ commutative functions

$$f_m(x,y) = f_m(y,x); \quad x,y \in \{0,1,2,\ldots,b-1\}; \quad m \in \{0,\ldots,\ell-1\}$$

The key k_{ij} for communication between user i and j is then defined by

$$k_{ij} = \sum_{m=0}^{\ell-1} f_m(a_{im}, a_{jm}).$$

Here we assume that the functions $f_m(.,.)$ have subsets of $GF(2^M)$ as their respective range of values

We do not assume that the functions $f_m(.,.)$ have any other property than commutativity. Hence we consider them to be tables. In calculating keys k_{ij}, user i always uses $f_m(a_{im}, .)$ and thus he only has to store b values for each function. Therefore the number of bits that each user has to store is

$$S_{C1} = M\ell b = (b/\log b) \cdot M \cdot \log N$$

which is minimized for b=3. However, if two or more users combine their information about the functions $f_m(\cdot,\cdot)$ we see that they can always generate extra keys.

To counter the threat that a small group of users combine their knowledge of the functions $f_m(\cdot,\cdot)$ and generate new keys, we use a linear code to generate extended addressvectors $\underline{a}'_i$ from $\underline{a}_i$. $\underline{a}'_i$ is then used in the same way to generate the key.

We denote the dimension of $\underline{a}'_i$ with $n>\ell$. We also assume b to be a prime number which enables us to perform calculations in GF(b). Let G denote the $\ell \times n$ generating matrix of the code, then $\underline{a}'_i = \underline{a}_i G$. We want G to be a maximum distance separable (MDS) code[1]. The property of MDS codes we use is that every ℓ columns of G are linearly independent. This means that any ℓ elements in a codeword $\underline{a}'_i$ uniquely determines $\underline{a}_i$. Hence two codewords can at most have $\ell-1$ elements in common and we see that to generate an "extra" key, at least $\lceil n/(\ell-1) \rceil$ users have to cooperate.

MDS codes do not exist for all values of ℓ, n and b. However, it is known that with b a prime there always exist a code with n satisfying the following relation.

$$n \leq \begin{cases} b+1 & \text{if } 2 \leq \ell \leq b \\ \ell+1 & \text{if } b < \ell \end{cases}$$

As we want $n/(\ell-1)$ to be large we can assume that $b+1 \geq n > \ell$. We shall also use t, the number of users that have to cooperate to generate "extra" keys, as a design parameter. Let $n = (t-1)(\ell-1)+1$ then the storage requirements for this schem becomes $S_{C2} = M \cdot n \cdot b = M((t-1)(\ell-1)+1) \cdot b$. To see what this means we upper bound the minimum of S_{C2}. The conditions used are $b^\ell = N$ and $(t-1)(\ell-1)+1 \leq b+1$ which leads to (when b>3) $S_{C2} < M(t-1)^2 (\log N)^2$.

A SCHEME BASED ON POLYNOMIALS IN 2 VARIABLES

The idea behing this scheme is to use a polynomial $p(x,y)$ in GF(q), having the property that $p(x,y) = p(y,x)$. For this scheme we assume that each user is associated with an unique element i in GF(q) and we use this i to identify the user. We also assume that q is in the order of 2^M, that is we represent the elements in FG(q) with M bits. To generate a key for users i and j one evaluates $p(i,j)$. We observe that a specific user i only has to know the polynomial $p(i,y)$. Hence each user knows only a part of the total polynomial. We define $p(x,y)$ by

$$p(x,y) = (x^0, x^1, \ldots, x^{n-1}) A \ (y^0, y^1, \ldots, y^{n-1})^T$$

where A is a symmetric n×n matrix.

We see that each user only has to store n coefficients namely the vector

$$\underline{b}_i = (i^0, i^1, \ldots, i^{n-1}) A.$$

The calculation of a key k_{ij} then consists of first calculating $(j^0, j^1, \ldots, j^{n-1})$ and then performing scalar multiplication of this vector and $\underline{b}_i$.

Observe that if $q > n$ (in our case $q \gg n$) then every n distinct vectors $(j^0, j^1, \ldots, j^{n-1})$ are linearly independent. This can be seen by observing that if n different vectors are taken as rows in a n×n matrix, the matrix will be a Vandermonde matrix, which will have a nonzero determinant.

The security against cooperating users can now be evaluated as follows: A contains $n(n+1)/2$ independent elements. Each user knows a set of n coefficients $\underline{b}$, i.e. the values of n equations. Hence to completely solve for the elements of A we need at least $\lceil n(n+1)/(2n) \rceil = \lceil (n+1)/2 \rceil$ vectors $\underline{b}$. To guarantee that at least n users have to cooperate to generate an "extra" key, the least possible is $n = 2(t-1)$ and then $S_p = M \cdot 2(t-1)$.

We also observe that when $\lceil (n+1)/2 \rceil$ users cooperate they can solve for A and then they can generate all keys in the system. However, up to this point, when the system totally breaks down, no extra information at all is gained about the value of a specific "extra" key.

CONCLUSIONS

We have shown two different key generation schemes that require small secret storage at the users while at the same time having simple functions for calculation of legal keys. The first scheme based on MDS codes is good for situations when there is no need to protect the key scheme against large groups of cooperating users trying to generate extra keys. The second scheme can handle such situations. However, when enough users cooperate and succed to generate one extra key in the polynomial based system they can generate all keys in the system. It would be nice to have systems that degrade more gracefully but here more research is needed.

REFERENCE

1. F. J. MacWilliams and N. J. A. Sloane, chapter 11 in: "The Theory of Error Correcting Codes", North Holland, Amsterdam (Third Printing 1981).

CRYPTOGRAPHIC SOLUTION TO A MULTILEVEL SECURITY PROBLEM

Selim G. Akl and Peter D. Taylor

Queen's University
Kingston, Ontario
Canada

ABSTRACT

A scheme based on cryptography is proposed for enforcing multilevel security in a system where hierarchy is represented by a partially ordered set (or poset). Straightforward implementation of the scheme requires users highly placed in the hierarchy to store a large number of cryptographic keys. A time-versus-storage trade-off is then described for addressing this key management problem.

1. INTRODUCTION

Assume that the users of a computer (or communication) system are divided into a number of disjoint sets, $U_1, U_2, \ldots, U_n$. The term <u>security class</u> (or <u>class</u>, for short) will be used to designate each of the U_i. Assume further that a binary relation $\leq$ partially orders the set $S = \{U_1, U_2, \ldots, U_n\}$ of classes. The meaning of $U_i \leq U_j$ in the partially ordered set $(S, \leq)$ is that users in U_i have a <u>security clearance</u> lower than or equal to those in U_j. Simply put, this means that users in U_j can have access to information held by

---

This work was supported by the Natural Sciences and Engineering Research Council of Canada under Strategic Grant G0381.

(or destined to) users in U_i, while the opposite is not allowed.

Let x_m be a piece of information, or **object**, that a central authority (CA) desires to store in (or broadcast over) the system. The meaning of the subscript m is that object x is accessible to users in class U_m. The partial order on S implies that x_m is also accessible to users in all classes U_i such that $U_m \leq U_i$. It is required to design a system which, in addition to satisfying the above conditions, ensures that access to information is as decentralized as possible. This means that authorized users should be able to <u>independently retrieve</u> x_m <u>as soon as it is stored or broadcast</u> by CA.

This <u>multilevel security problem</u> arises in organizations where a hierarchical structure exists. Government, diplomacy and the military are examples of such hierarchies. Applications also exist in business and in other areas of the private sector, for example in the management of databases containing sensitive information, or in the protection of industrial secrets. Finally, the model is used in the design of computer operating systems to control information flow from one program to another.[1,2]

This paper presents a solution to the multilevel security problem based on cryptography. Familiarity with modern cryptology will henceforth be assumed.[2,3,4]

2. CRYPTOGRAPHIC SOLUTION

In the following we will assume the presence of a cryptoalgorithm with <u>enciphering procedure</u> E to be used under the control of an <u>enciphering key</u> K^e. The notation

$$u = E_{K^e}(v)$$

means that u is the result of enciphering v using E and K^e. A <u>deciphering procedure</u> D and <u>deciphering key</u> K^d are used to recover v:

$$v = D_{K^d}(u).$$

If enciphering and deciphering are symmetric,[5] then K^d is identical to (or can be easily obtained from) K^e. If the cryptoalgorithm is of the asymmetric type,[5] on the other hand, K^e and K^d are different and knowledge of K^e is of no help in computing K^d. The superscripts e and d will henceforth be omitted as K^e is only used with E and K^d with D.

A step-by-step description of a cryptographic solution to the multilevel security problem is given below.

Step 1: The central authority generates n (deciphering) keys, $K_1, K_2, \ldots, K_n$, for use with the cryptoalgorithm.

Step 2: For $i = 1, 2, \ldots n$, key K_i is distributed to all users in U_i who keep it secret.

Step 3: In addition, for $i, j = 1, 2, \ldots, n$, all users in U_j also obtain K_i if $U_i \leq U_j$.

When an object x_m is to be stored (or broadcast) it is first encrypted with K_m to obtain:

$$x' = E_{K_m}(x_m)$$

and then stored (or broadcast) as the pair $[x', m]$. This guarantees that only users in possession of K_m will be able to retrieve x_m from:

$$x_m = D_{K_m}(x'). \qquad \square$$

As an example, Fig. 1 shows the Hasse diagram of a poset $(S, \leq)$ with $n = 9$; the keys held by each class in S will be as follows:

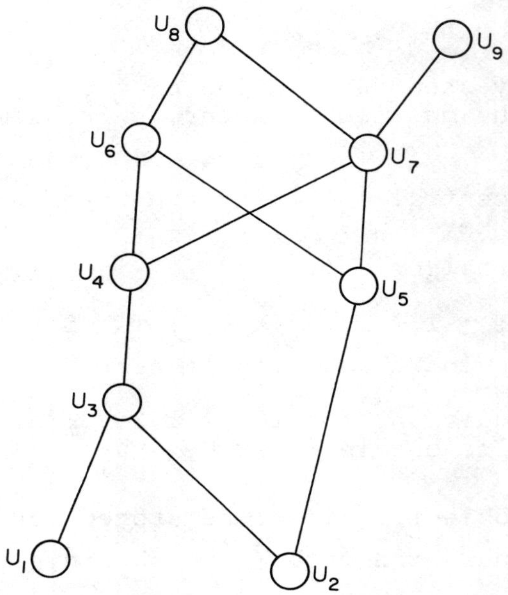

Fig. 1 Hasse diagram of the poset $(\{U_1, U_2, \ldots, U_n\}, \leq)$. The meaning of $U_i \leq U_j$ is that U_i has a security clearance lower than that of U_j.

Class	Keys
U_1	K_1
U_2	K_2
U_3	K_1, K_2, K_3
U_4	K_1, K_2, K_3, K_4
U_5	K_2, K_5
U_6	$K_1, K_2, K_3, K_4, K_5, K_6$
U_7	$K_1, K_2, K_3, K_4, K_5, K_7$
U_8	$K_1, K_2, K_3, K_4, K_5, K_6, K_7, K_8$
U_9	$K_1, K_2, K_3, K_4, K_5, K_7, K_9$

This solution has the advantage that only one copy of x_m is stored or broadcast. Similarly, the operations of enciphering and deciphering are performed exactly once. Its disadvantage is the large number of keys held by each user. The worst case occurs when, for some j, $U_i \leq U_j$ for all i, and users in U_j have to store n keys. The following section deals with this problem.

3. KEY MANAGEMENT

We solve the key storage problem by proposing a system in which a user in U_j stores only his own key K_j, and can compute from this the key K_i if and only if $U_i \leq U_j$. Formally, whenever $U_i \leq U_j$ we define computable functions g_{ij} for which $K_i = g_{ij}(K_j)$, but if $U_i \not\leq U_j$, K_i cannot be computed from K_j. We first look at the case of a totally ordered set.

3.1 Totally Ordered Sets

The simplest case of a partially ordered set occurs when the set is totally ordered, i.e. when $U_1 \leq U_2 \leq \ldots \leq U_n$. Suppose f is a function which is one-way, and indeed for which f^m (where the power denotes composition) is one-way for all positive integers m. Here by one-way we mean easy to compute but extremely difficult to invert from a computational point-of-view. Then if K_n is randomly selected, we define

$$K_{n-1} = f(K_n)$$

$$K_{n-2} = f(K_{n-1})$$

.

.

.

$$K_1 = f(K_2)$$

Thus, $g_{ij} = f^{j-i}$. Since positive powers of f are one-way, g_{ij} can only be computed if i is not greater than j.

As an example, with any cryptoalgorithm, we could set $f(K) = E_k(K)$ where k may or may not depend on K. We will present two examples of this type, a symmetric cryptosystem, for which we set k=K in order to make f one-way, and an asymmetric one for which k is independent of K.

3.1.1 k=K.
If E_K is the enciphering procedure for any cryptoalgorithm then $f(K) = E_K(K)$ is a candidate for our one-way function f. Then our key transformation is $K_{i-1} = E_{K_i}(K_i)$. For example if the Data Encryption Standard (DES)[6] is used, then K_i must be a 56-bit key. The only way we can see to invert f is with an exhaustive cryptanalytic attack using 2^{56} (or on the average 2^{55}) steps. This remote threat can be easily made more improbable by asking each user in U_i to store two keys K_i and k_i and computing K_{i-1} and k_{i-1} from

$$K_{i-1} = E_{K_i}(k_i)$$

and $k_{i-1} = E_{k_i}(K_i)$ for i = n,n-1,...,2.

The average effort required by a user in U_m to go just

one level up in the hierarchy to find the key for U_{m+1} is now 2^{110} steps!

3.1.2 k independent of K.
If the CA generates two large primes p and q (of the order of 100 digits each) and makes their product M=pq public, then the scheme $f(K) = K^2 \pmod{M}$ appears to have the required properties. The key transformation equation reads $K_{i-1} = K_i^2 \pmod{M}$ and is believed to be difficult to invert without knowledge of (the Euler totient function) $\phi(M)$ which is equivalent to factoring M.[7,8,9]

3.2 Arbitrary Posets

Our solution for arbitrary posets has the following form. We assign an integer t_i to each class U_i so that

$$U_i \leq U_j \text{ if and only if } t_j | t_i.$$

An example of such an assignment appears in Fig. 2 with the t_i recorded in the U_i node. We then define a family $\{f_m\}$ of one-way functions with the property $f_{m \cdot z} = f_m * f_z$ where m and z are integers and the * denotes composition. The CA chooses a random K_o and defines

$$K_i = f_{t_i}(K_o).$$

Then if $U_i \leq U_j$, $t_i = z\, t_j$ for some integer z, and

$$K_i = f_{t_i}(K_o) = f_z * f_{t_j}(K_o) = f_z(K_j)$$

so that K_i can be computed from K_j and $g_{ij} = f_{t_i/t_j}$. The trick is to choose the f_m so this calculation can be made only when $U_i \leq U_j$.

This can be done with a generalization of 3.1.2. The CA chooses two large primes p and q and makes M=pq public. Then

$$f_m(K) = K^m \pmod{M}.$$

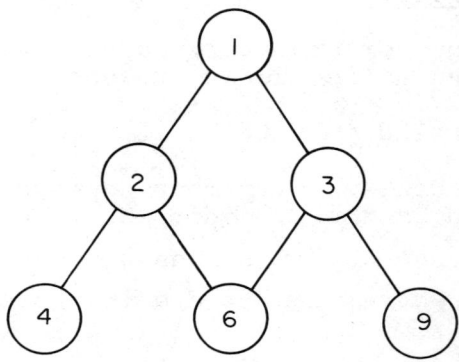

Fig. 2 An integer t_i is assigned to each U_i such that $U_i \leq U_j$ if and only if $t_j | t_i$. Key K_i is obtained from key K_j using the transformation $K_i = [K_j]^{t_i/t_j} \mod M$, where M is the product of two large primes.

The key transformation equation becomes

$$K_i = K_o^{t_i} = \left[K_o^{t_j}\right]^{t_i/t_j} = [K_j]^{t_i/t_j} \pmod{M}$$

which is apparently computable if and only if $t_j | t_i$. Indeed if t_i/t_j is not an integer, the computation requires extraction of a root (mod M) and this seems to be difficult.

Finally we suggest an algorithm for the assignment $\{t_i\}$. Each class U_i is assigned a distinct prime p_i and

$$t_i = \prod_{U_j \not\leq U_i} p_j ,$$

where, by convention, an empty product equals 1. This is illustrated in Fig. 3, with the p_i underneath the node, and t_i inside. This assignment is not nearly as efficient as the ad hoc assignment of Fig. 2, but it is systematic and easily programmed. To verify that this assignment works, we must show that $t_j | t_i$ if and only if $U_i \leq U_j$. First if $U_i \leq U_j$ then the product defining t_j has less terms than that for t_i, and $t_j | t_i$. Conversely if $U_i \not\leq U_j$ than $p_i | t_j$ (by definition of t_j) which implies t_j cannot divide t_i (since p_i does not divide t_i).

3.3 A Symmetric Cryptoalgorithm Based On Modular Exponentiation

We now suggest a cryptoalgorithm which may work well with the key management system presented in 3.2. The CA chooses a large prime N (larger than the number M of 3.2) and makes it public. Each class U_i receives a private number b_i relatively prime to N-1. Every object to be enciphered is broken into blocks each of which is expressed as an integer x smaller than N. Enciphering and deciphering procedures are given by

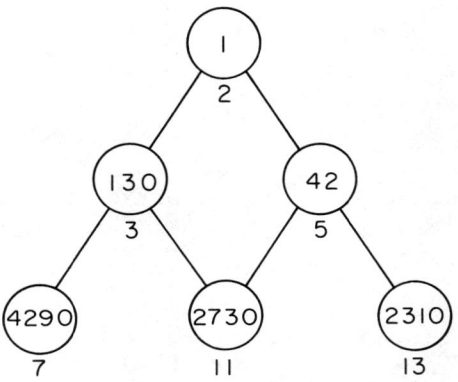

Fig. 3 Each class U_i is assigned a distinct prime p_i (shown underneath the nodes). The t_i's (shown inside the nodes) are obtained from

$$t_i = \prod_{U_j \not\subseteq U_i} p_j.$$

$$x' = (x)^{a_i} \mod N$$

and

$$x = (x')^{b_i} \mod N,$$

respectively. Here a_i is chosen so $a_i b_i = 1 \mod (N-1)$ for all i. The cryptoalgorithm is symmetric for, given N, a_i and b_i can easily be derived from each other.

This scheme could be used with the key management system of 3.2 with $b_i = K_i$, provided K_i is relatively prime to N-1. When this fails, an agreed upon algorithm, common to all users, can be used to generate from K_i a number b_i relatively prime to N-1. Such an algorithm is presented in the Appendix.

ACKNOWLEDGEMENTS

Many thanks are due to Glenn MacEwen for bringing this problem to our attention. We are also indebted to Henk Meijer and Ibrahim Assem for many insightful discussions. Greg Wilson implemented the algorithm in the appendix and offered a number of useful comments.

REFERENCES

1. G. H. MacEwen, Secure information flow in distributed systems, Proceedings of the Eleventh Biennial Symposium on Communications, Queen's University, Kingston, Ontario (1982).
2. D. E. R. Denning, "Cryptography And Data Security", Addison-Wesley, Reading, Massachusetts (1982).
3. A. G. Konheim, "Cryptography: A Primer", John Wiley & Sons, Toronto (1981).
4. C. Meyer and S. M. Matyas, "Cryptography - A New Dimension In Computer Security", John Wiley & Sons, New York (1982).
5. G. J. Simmons, Symmetric and asymmetric encryption, Comput. Surv. 11: 305 (1977).
6. Data Encryption Standard, Federal Information Processing Standard (FIPS), Publication 46, National Bureau of Standards, U.S. Department of

Commerce (1977).
7. R. L. Rivest, A. Shamir and L. Adleman, A method for obtaining digital signatures and public-key cryptosystems, <u>Commun. ACM</u> 21:120 (1978).
8. M. O. Rabin, Digitalized signatures and public-key functions as intractable as factorization, Technical Report MIT/LCS/TR-212, Laboratory for Computer Science, Massachusetts Institute of Technology, Cambridge, Massachusetts (1979).
9. H. C. Williams, A modification of the RSA public-key encryption procedure, <u>IEEE Trans. Inform. Theory</u> IT26:726 (1980).

APPENDIX

Given two positive integers r and s, where s is even and r < s, the function 'relative_prime' returns a unique positive integer which depends on r and is relatively prime to s.

```
relative_prime (r,s)
   if (r is even) then r ← r+1
   outer ← false
      while (outer = false) do
         begin
            w ← r
            inner ← false
            while (inner = false) do
               begin
                  x ← gcd (w,s)
                  if (x = 1) then inner ← true
                  else w ← w/x
               end
            if w ≠ 1 then outer ← true
            else r ← r + 2
         end
relative_prime ← w.
```

LOCAL NETWORK CRYPTOSYSTEM ARCHITECTURE:
ACCESS CONTROL

Thomas A. Berson

Sytek, Inc.
1225 Charleston Road
Mountain View, CA 94043

INTRODUCTION

A mechanism which uses keyed verification sequences at the link layer of protocol to control access to data transport facilities of a local area network will be described.

The local network itself has been described previously [1]. For purposes of this discussion, its salient features are these: the transmission plant has the physical topology of a tree. Network nodes (users) are connected at some of the leaves. All transmissions from any node proceed up the tree to its root. At the root is equipment referred to as the Data Channel Access Module (DCAM). This rebroadcasts transmissions received at the root back down the tree toward all the leaves. The network itself contains no switches; the selection of appropriate data from the rebroadcast data stream is amongst the functions assigned to a microprocessor in each of the network nodes.

The transmission plant may be constructed of standard cable television (CATV) components. In fact, a single transmission plant may be easily shared between television program distribution and local area data network functions. It is envisioned that such shared networks will be widely installed in metropolitan areas. A single network in a metropolitan environment might reasonable be expected to have many tens of thousands of leaves within an area of several thousand square kilometers.

While it is not practical to restrict or control physical access to so widespread a network, some sort of access control mechanism must be provided if the owner of the network is to derive revenue from the sale of its data transport facilities to network subscribers. The link level scheme presented here offers one solution to this problem.

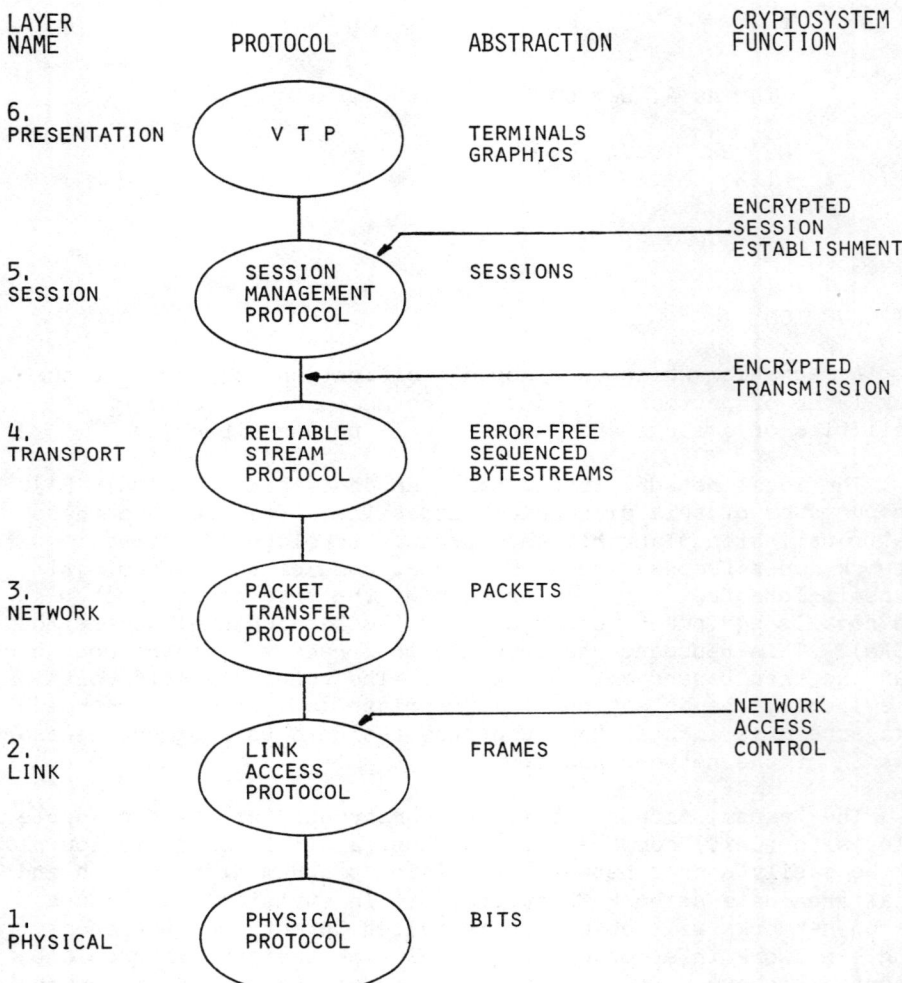

Figure 1 -- The protocol architecture. Showing layer names and numbers in correspondence with the OSI model. Also showing protocol names, the abstraction created by each layer, and the cryptosystem responsibilities of each layer.

PROTOCOL LAYERING

Nodes on the network establish, control and carry out communication by means of conventions known as the network protocols. These may be thought of as a hierarchy of abstract machines, each responsible for some aspect of the communication, and each implemented using the facilities made available by protocol machines at lower layers.

Figure 1. illustrates the layering of protocols in the present network. Other networks may vary slightly in the details of various layers, but the order of the layers themselves is becoming standardized under the influence of the International Standards Organization which refers to this particular hierarchy as the Open Systems Interconnection Reference Model. Figure 1 shows the name of each layer, the name of the protocol at that layer, and the abstraction implemented by that protocol. Also shown are the placement of network cryptosystem functions.

The network supports an end-to-end cryptosystem at the session level [2]. This protects user data against passive and some active channel tapping attacks. Keys for this system are distributed as part of the session layer protocol. the encryption takes place just "above" the transport layer.

The layer of concern to the present access control scheme is the link layer, which is far more primitive than the layers involved in the end-to-end cryptosystem.

THE LINK LAYER

The link layer lies just above the lowest layer of protocol, called the physical layer. The physical layer controls the electrical conventions for establishing a bit and its value and provides for the transmission of bits. The link layer deals with the transmission of collections of bits referred to as frames. The link layer's traditional functions include packaging bits into frames, data transparency, node addressing, and error detection. To these, we add an access control function.

A link layer frame is shown schematically as Figure 2. Each frame contains within it the units of information of each higher protocol layer which is active. These are shown in Figure 2 as a set of nested parentheses and will not be of concern to our access control scheme. Also shown are a source address S, a destination address D, a frame verifier FV, and an error detection code CRC. These S, D, FV and CRC are illustrative of header information at the link layer. They and other data items required for proper operation of the link layer are transmitted with each frame.

Figure 2 -- A frame, the unit of information at the link layer. The nested parentheses represent units of information at higher layers. The source address S, destination address D, frame verifier FV, and error detecting code CRC are illustrative of link layer header fields.

ACCESS CONTROL

The access control scheme is comprised of policy and mechanism. These are independent and reside on different network nodes. The policy decisions are made in one or more specialized policy server nodes. Any node wishing to use the network must first request access from the policy server. If the request is granted, the policy server securely distributes a key variable to both the requesting node and the DCAM. This variable is used by each recipient to generate identical pseudorandom sequences. Successive words from this sequence are used by the requesting node as frame verifiers for transmitted frames. The DCAM checks all incoming frames, and rebroadcasts only those whose frame verifiers are the "next" word from the transmitting node's pseudorandom sequence.

ACCESS CONTROL MECHANISM

Mechanization of this scheme requires:

a. That every frame contain a frame verifier (FV).

b. That the DCAM only rebroadcasts frames which have "valid" frame verifiers.

c. That there be a specialized node, the Network Access Controller (NAC), on the network which can make access policy decisions.

d. That there be a private channel between the DCAM and the NAC.

e. That every node contain a secret Master Access Key (MAK).

f. That the NAC have a record of the MAK of each node which might be granted access.

Local Network Cryptosystem Architecture: Access Control

The mechanism operates as follows, refer to Figure 3.

1. A node wishing channel access transmits a frame containing the request. This frame does not have a valid frame verifier so it is not rebroadcast by the DCAM.

2. Because the frame contains a channel access request it is sent via the private channel to the NAC. (The private channel might be an out-of-band circuit or it might be an end-to-end encrypted session on the network itself.)

3. The NAC makes its policy decision. This for example might be based upon the status of the requesting node's account, or upon the time of day, etc.

4a. If the request is denied, a frame containing the reason for denial is sent to the requesting node. This frame contains a valid FV from the NAC's sequence and so is passed by the DCAM. If the request has been denied no further action occurs.

4b. If the request is granted, a frame containing the grant is sent to the requesting node. The grant includes a key created for the purpose of the grant and known as the channel access key (CAK). This is encrypted in the master access key of the requesting node. The frame containing the grant is passed by the DCAM as it contains a valid FV from the NAC's sequence.

5. The CAK and requesting node ID are sent from the NAC to the DCAM over the private channel. The DCAM adds this information to its list of currently authorized network users.

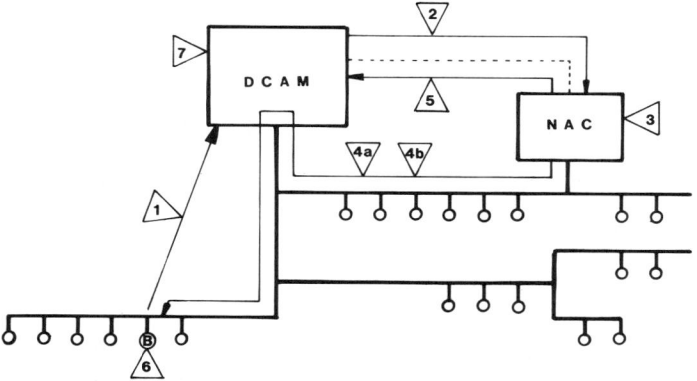

Figure 3 -- The mechanism by which access is granted to requesting node "B" involves an exchange of messages between B, the DCAM, and the NAC.

6. The requesting node decrypts the grant to recover the CAK. This CAK is then used to generate a pseudorandom sequence. Words from that sequence are taken in order and used as FVs in succeeding transmitted frames.

7. The DCAM generates the same pseudorandom sequence from the CAK. The FV in every frame received from a node is compared with the next unused word of that node's sequence. The frame is rebroadcast only if the received FV and the locally computed FV are identical.

SEQUENCE GENERATION

The cryptographic requirement on the pseudorandom sequence from which frame verifiers are taken is that it be unpredictable without knowledge of the key. That is to say that perfect knowledge of the means of sequence generation and of some portions of the sequence should be of no aid in predicting other portions of the sequence.

There are a number of sequence generators which meet this requirement. For example, DES [3] can be used in the "counter mode" with a 120-bit key variable. 56 of these bits form the DES Key and 64 bits are used as in initial plaintext block (IV). IV is encrypted to form the first 64 bits of the pseudorandom sequence. Subsequent blocks of the sequence are formed by encrypting IV+1, IV+2, ... Frame verifiers are formed by regrouping the sequence into words of appropriate length, w.

Another classical way to generate a suitable sequence is to use shift registers. Linear shift registers can be combined with nonlinear components to form sequence generators with known properties as described by Beker [4]. A third interesting possibility is to use an x^2 mod N generator as described by Blum, Blum and Shub [5].

An implementation option is to use a "random access" pseudorandom sequence -- one in which the i'th word can be directly computed from i and the key without first requiring computation of all previous i-1 words. This feature might be useful in engineering the DCAM as it allows memory size vs. processing power tradeoff. The DES generator and that of the Blums are random access sequences. That of Beker is not.

ATTACKING THE SCHEME

Consider the problem of an attacker who wishes to sneak one unauthorized frame through the DCAM. We will show that he can expect to succeed in 2^{w-1} trials, where w is the FV word size in bits.

Assume that the FV are uniformly distributed in $[0, 2^w-1]$. That is, for all keys and for all i, the probability that FV_{ki} equals some constant in the range $[0, 2^w-1]$ is 2^{-w}.

Assume further that the attacker knows the identity A of a node whose use of the network is currently authorized, but which, for sake of simplicity, is not active at the moment. This can be easily determined by monitoring the network traffic. The attacker does not know A's current CAK.

The attacker begins to send frames with A as the sender's identity and with FV=c, where c is some arbitrary constant in the range of possible FVs. The DCAM will at first expect FV_i, then FV_{i+1}, etc. from A's sequence. Because the FV are uniform, only 2^{w-1} trials are required on average before the transmitted c equals the expected FV and the frame is passed as authentic.

An obvious (but ineffective) countermeasure is for the DCAM to not entertain any further frames from a node where some $1 \leq t \ll 2^{w-1}$ consecutive frames have not had valid FVs. To counter this countermeasure the attacker collects $(2^w)/t$ authorized but inactive sender identities A_j. He then attacks as before, holding the FV constant at c but changing his claimed identity to a new A_j every t trials. He will again succeed on average after 2^{w-1} trials.

SUMMARY

We have described how keyed pseudorandom sequences may be used at the link layer of protocol to control access to the transport facilities of a network. We have discussed one mechanization of such a scheme -- there are no doubt many other equivalent mechanizations. We have also indicated a variety of useful sequence generators. Last, we have shown that the scheme is not perfect, and that its strength depends upon the length of the frame verifiers. Choice of this length is a tradeoff between nonrevenue "overhead" bits in the frame header and nonrevenue "stolen" bits in unauthorized frames. Similarly, for any given network the choice of a mechanization and of a sequence generator will be driven by the network's engineering and economic constraints.

REFERENCES

[1] K.J. Biba, "LocalNet(tm): a digital communications network for broadband coaxial cable," Proc. IEEE CompCon, (Feb. 1981), pp. 59-63.

[2] T.A. Berson and R.K. Bauer, "Local network cryptosystem architecture," Proc. IEEE CompCon, (Feb. 1982), pp. 138-143.

[3] <u>Data Encryption Standard</u>, FIPS PUB 46, U.S. Department of Commerce, National Bureau of Standards, (January, 1977).

[4] H. Beker and F. Piper, <u>Cipher Systems</u>, Northwood Books, London (1982).

[5] L. Blum, M. Blum and M. Shub, "A simple secure pseudo-random number generator," Memorandum UCB/ERL M82/65, Electronics Research Laboratory, College of Engineering, University of California, Berkeley (Sept. 1982).

IMPLEMENTING AN ELECTRONIC NOTARY PUBLIC

Leonard M. Adleman

University of Southern California
Massachusetts Institute of Technology

ABSTRACT

Many communication security problems admit both "physical" and "mathematical" solutions. For example sending a message from A to B without exposing it to C, can be accomplished physically by means of secure courier, or mathematically by means of encryption. With the advent of public key cryptography, many problems originally believed to be solvable only by physical means have been shown to have mathematical solutions (e.g. key distribution [DH], secret sharing [S], coin flipping [B], mental poker playing [SRA]). In this paper we describe a mathematical solution to a communication security problem, which arose in connection with the Nuclear Test Ban Treaty, and for which only physical solutions were known. The problem concerns the implementation of an electronic notary public - a device which can certify information for a group of mutually distrusting parties - among which may be builder of the device.

A NOTARY PUBLIC

The function of a notary public is to ceritfy that an "event" took place at a particular time and place. For example that on January 1, 1982, in Washington, D.C., a man calling himself Fred Smith signed a document which read "I Fred Smith leave all my worldly possessions to John Brown."

The function of the notary public is well filled when all parties concerned with the event are sure that the notary public is trustworthy, and in fact he is. Problems arise when the notary public may have a vested interest in the event - for example if his name is John Brown.

With the advent of public key cryptosystems [DH] [SRA] and digital signature it has become conceivable to replace the human notary public by an electronic one. At first this appears to remove any questions of trustworthiness. In fact, it does not. Ultimately parties concerned with a notarized event, must be sure that the device itself does not contain a "trojan horse" which embodies the vested interest of its builder. In this paper we describe a method for overcoming this problem, under the assumption that factoring integers is difficult.

THE LEAKPROOF ASSUMPTION

It is important to clarify what kind of probelm we are solving and what kind we are not.

Consider the classical problem of sending a message from A to B without revealing it to C. The standard physical solution is to use a courier. The standard mathematical solution is to encrypt - lets say using the RSA system.

In the real world the encryption solution's success depends on implementation. Failures can arise for the following reasons at least:

1. <u>Leaks</u>. The device used to encrypt physically leak information which is valuable to C. For example C built the device and added an extra radio transmitter to feed him the secret key. Or the device (either by C's doing or by bad design) alters the strength or timing or some other physical parameter of the "legitimate" outputs in a way which gives important cryptanalytic information to C.

2. <u>Trojan Horses</u>. C manipulates the algorithms of the encryption and decryption devices so that they behave in a manner benefiting C. For example by modifying the algorithms to always generate an encryption and decryption key pair which C has previously decided on. Then when A sends the message to B using the devices, C is in a position to intercept.

While the particular trojan horse in our example is easy to overcome, it illustrates the type of problem we are trying to solve in this paper - removing all conceivable trojan horses from an electronic notary public. These problems are algorithmic and subject to mathematical solution. Problems of leaks are physical. They are the concern of physicists and engineers (in fact agencies like the NSA invest great resources on just such concerns). However, they do not admit mathematical solution and do not concern us in this paper.

THE NUCLEAR TEST BAN TREATY PROBLEM*

As part of the Nuclear Test Ban Treaty, Russia has agreed to allow the United States to implant seismographic devices in Russian soil. The devices will monitor earth tremors and thereby detect nuclear activity. The devices will have radio transmitters which will beam the information to the U.S. via satellites. Apparently the U.S. is content that the devices can be made leakproof and tamperproof; however, concern about the integrity of the transmissions from the devices generated the following sequence of communications problems.

Problem 1

The U.S. wanted some method for protecting the transmissions from insertions and deletions, lest false transmissions indicating a halcyon state be sent while in fact testing is occurring.

Solution 1

Buried with the seismographic device is a U.S. built encryption device with key K. The transmissions are encoded with K, and when they reach the U.S. they are decoded with K.

Problem 2

The Russians wanted some method of monitoring transmissions leaving their country, lest "spy" information (e.g. weather condition) be transmitted along with the seismographic signal.

Solution 2

(Proposed by Gus Simmons of Sandia Laboratories) Use the RSA public key cryptosystem. The U.S. will generate a pair of keys K_E (for encoding) and K_D (for decoding), K_E will be buried with the seismographic device. K_D will be given to the Russians.

Problem 3A

The U.S. is concerned that if the Russians do test, and the transmissions reach the U.S. they will not be able to protest to the world community, since the Russians are in a position to claim that the U.S. (which generated K_E) fabricated the transmissions.

---

*The author was not privy to the actual US-USSR discussion of this problem. The problem described here is his understanding based on informal, unclassified discussions with some of the people involved.

Problem 3B

The Russians are concerned that even if they don't test, the U.S. may fabricate transmissions and condemn them in the world community.

Solution 3

The U.S. will build each device with a random number generator (e.g. geiger counter) in it which will generate a number to be used as a "seed" for generating K_E and K_D. While K_D will still be given to the U.S. and USSR, no one but the buried device will know K_E - not even the U.S. K_D will be made public (e.g. by publishing it in Pravda and the N.Y. Times and reciting it at the U.N.). Then any transmissions which are invertable with K_D will necessarily have actually come from the device (which now is acting like a notary public). No denials or fabrications are possible. (It should be noted that no third party could be found which both sides trusted to generated K_E and build the devices).

Problem 4

The Russians would prefer to build the device for fear that the U.S. will put in a trojan horse which will generate a predetermined K_E.

Solution 4

The U.S. will build 6 devices and the Russians will choose one to be buried and keep the other 5 for "postmortum" examination to see if they can detect a trojan horse.

Problem 5

See problem 4.

Solution 4 is a "physical" solution. In the next section we will give a mathematical one. The interested reader may find it entertaining to consider possible solutions before turning to the next section.

PROTOCOL FOR IMPLEMENTING AN ELECTRONIC NOTARY PUBLIC (ENP)

In this section we give a general protocol for implementing an electronic notary public. The protocol can easily be adapted to solve the Nuclear Test Ban Treaty problem.

It is assumed, that the following agreements have been made.

1. A (for Alice) and B (for Boris) agree that A will build the ENP.

2. A and B agree on an "intended algorithm".* That is, an algorithm which A is supposed to implement in the ENP, but which B is worried may in fact be replaced by a trojan horse.

3. B agrees that if the intended algorithms really is implemented in the ENP then he is satisfied.

4. (Optional) A and B agree that the intended algorithm will have a "key generation" phase during which A and B can make inputs (perhaps none) for the purposes of generating an encryption key K_E and a decryption key K_D of a public key cryptosystem. K_D will be output to both A and B and K_E will forever remain in the (leakproof) ENP. After the key generation phase K_E will be used to certify information through a port set up for that purpose.

Some reflection will show the following requirements must be satisfied.

R1. At the end of the key generation phase, A and B must each have sufficient information to be sure that if the intended algorithm had been implemented then K_D would have been the output.

R2. At the end of the key generation phase, A and B must each be sure that the other <u>could not</u> have sufficient information to determine K_E.

R1 is a strong requirement. If R1 is met and the device does not output K_D or ever notarizes an event by a signature different from one which K_D inverts, then B can detect immediately that the intended algorithm is not being used and that A has substituted a trojan horse.

If R2 did not hold for A and B then of course which ever one it did not hold for would be able to forge notarizations.

One serious problem in meeting requirements R1 and R2 is that all known algorithms for generating public key cryptosystem keys generate them "simultaneously". That is, from common information (a "seed"). This seems to be an intrinsic property of public key systems. In our solution to the ENP problem our intended algorithm will use an RSA system which in the usual way generates

---

*To be formally correct "intended algorithmic function" would be more appropriate.

both K_E and K_D from such a common "seed". Thus to satisfy R1 and R2 we must create a situation where despite this property of the intended algorithm, and the fact that both A and B possess copies of it, neither A nor B can figure our which K_E it will generate, even though they can both figure out which K_D it will generate.

The Protocol

1. A produces two primes P_A, P_A' (of say 500 bits each) and a random number R_A (of 1000 bits).

2. Analogously B produces P_B, P_B', and R_B.

3. A gives R_A and $N_A = P_A P_A'$ to B

4. B gives R_B and $N_B = P_B P_B'$ to A

5. A inputs (without revealing to B) P_A, P_A', R_A to the ENP.

6. B inputs (without revealing to A) P_B, P_B', R_B to the ENP.

The intended algorithm is:

"On input P_A, P_A', R_A, P_B, P_B', R_B.

Calculate $R = R_A \oplus R_B$

Calculate $PHI = (P_A-1)(P_A'-1)(P_B-1)(P_B'-1)$

Calculate $S = R^{-1} MOD(PHI)$

Calculate $N = P_A \cdot P_A' \cdot P_B \cdot P_B'$

Output $K_D = (R,N)$

Use $K_E = (S,N)$ for notarization"

R1

Since A and B each have N_A, N_B, R_A, R_B each can calculate $N = N_A \cdot N_B$, $R = R_A \oplus R_B$ and therefore knows what $K_D = (R,N)$ the intended algorithm would produce.

R2

Assume A knows $K_E = (S,N)$ then A knows $R \cdot S \equiv 1 \, MOD(PHI)$ from which using Millers results [M] it follows that A can factor N into P_A, P_A', P_B, P_B'; therefore, A can factor N_B which contradicts the assumption that factoring is intractable. Similarly for B.

REFERENCES

[B]   Blum, R., "How to Exchange (Secret Keys)" ERL Technical Memo UCB/ERL M81/90.
[DH]   Diffie, W. and Hellman, M., "New Directions in Cryptography," IEEE Trans. Inform. Theory, Vol. IT-22, Nov. 1976.
[M]   Miller, G.L., "Riemann's Hypothesis and Tests for Primality," Proc. Seventh Annual ACM Symp. on the Theory of Computing. Albuquerque, New Mexico, May 1975, pp. 234-239; extended vers. available as Res. Rep. CS-75-27, Dept. of Computer Science, University of Waterloo, Waterloo, Ont., Canada, Oct. 1975.
[RSA]   Rivest, R.L., Shamir, A. and Adleman, L.M., "A Method for Obtaining Digital Signatures and Public-Key Cryptosystems," CACM 21, pp. 120-126, February 1978.
[S]   Shamir, A., "How to Share a Secret," CACM 22, Number 11, p. 612, November 1979.
[SRA]   Shamir, A. Rivest, R. and Adleman, L.M., "Mental Poker," MIT/LCS/TM-125.

QUANTUM CRYPTOGRAPHY, OR
UNFORGEABLE SUBWAY TOKENS

Charles H. Bennett,[1] Gilles Brassard,[2]
Seth Breidbart[3] and Stephen Wiesner[4]

1. IBM Research, Yorktown Heights, NY 10598
2. Université de Montréal, Département d'I.R.O., C.P. 6128, Succ. "A", Montréal, Québec H3C 3J7
3. P.O. Box 1526, Wall Street Station, New York, NY 10268
4. MIT Research Laboratory of Electronics, MIT, Cambridge, MA 02139

ABSTRACT

The use of quantum mechanical systems, such as polarized photons, to record information gives rise to novel cryptographic phenomena, not achievable with classical recording media: 1) A Verify Only Memory (VOM) that, with high probability, cannot be read or copied by someone ignorant of its contents; 2) the multiplexing of two messages in such a way that, with high probability, either message may be recovered at the cost of irreversibly destroying the other.

Quantum multiplexing can be combined with public-key cryptography to produce unforgeable subway tokens that resist counterfeiting even by an opponent with a supply of good tokens and complete knowledge of the turnstiles that test them.

---

* Supported in part by the National Science Foundation and Canada's NSERC Grant number A4107.

INTRODUCTION

One of the first places public-key cryptography [1-2] was applied is at the Zero Power Plutonium Reactor in Idaho Falls, Idaho.[3] Because of the presence of fissionable materials, such as uranium and plutonium, it is important that only authorized persons be allowed in the facility. This is controlled by personalized access cards containing information on their bearers hand. The novelty about this scheme is that it includes a digitalized signature based on a trap-door one-way function. Because the computer that reads these access cards is not secure, the people who work at the facility could obtain the validation instructions. This would not enable them to forge cards for unauthorized persons, however, because of the asymmetry of public-key cryptography. (Notice that if the computer is indeed insecure, enemies might modify its programmation to introduce loopholes in the validation process.)

Security in the Idaho validation process depends on the fact that the access cards are personalized. Nothing prevents an enemy from copying cards that should fall into his hands, but of course such illegal copies would do him no good. We propose here unpersonalized access cards that cannot be reproduced. More precisely, it is infeasible for an enemy to come up with even a single counterfeit card that would allow him in the facility. This claim has to hold true if the would-be forger is allowed to perform any experiments whatsoever on any number of valid cards, and if he has complete knowledge of the validation algorithm. In short, anyone can verify if a given card is valid, yet only the mint can produce them.

Of course, there would be serious disadvantages to using unpersonalized access cards in high security areas: an enemy would be allowed in, sould he steal a valid card from an authorized person. For other applications, however, it would be unsuitable for the access cards to be personalized. This is the case, for instance, whenever authorization is available to the public at large upon payment of admission. Would it not be convenient for a transit authority to issue subway tokens that anyone could check for validity, yet no one could conterfeit? The impossibility of fraud should not depend on the use of a special type of paper or other similar conventional ideas that offer no real protection against well-equipped forgers, nor should it involve on-line communication with the transit authority. We propose here a scheme based on a fundamental idea of quantum physics: the impossibility of simultaneously determining rectilinear and diagonal polarization of photons.

ESSENTIAL PROPERTIES OF POLARIZED LIGHT [4]

Polarized light can be produced by sending ordinary light through a polarizing apparatus such as a Polaroid filter or Nicol prism. A beam of polarized light is characterized by its polarization axis, which is determined by the orientation of the polarizing apparatus in which the beam originates. Although polarization is a continuous variable, and can in principle be measured as accurately as desired by passing the beam through a second polarizer, the quantum mechanical uncertainty principle forbids measurements on any single photon from disclosing more than one bit about the beam's polarization. In particular, if a beam with polarization axis α is sent into a polarizer oriented at angle β, the individual photons behave dichotomously and probabilistically, being transmitted with probability $\cos^2(\alpha - \beta)$ and absorbed with the complementary probability $\sin^2(\alpha - \beta)$. Deterministic behaviour occurs only when the two axes are parallel (total transmission) or perpendicular (total absorbtion). If the two axes are not perpendicular, so that some photons are transmitted, one might hope to learn additional information about α by measuring the transmitted photons again with a polarizer oriented at some third angle; but this strategy is to no avail, because the transmitted photons in passing through the β polarizer, emerge with exactly β polarization, having lost all memory of their previous polarization α. Any other elementary two-state quantum system, such as a spin ½ atom, behaves similarly dichotomously and probabilistically.

VERIFY ONLY MEMORY

In order to get the reader used to quantum physics ideas, this section describes a very simple irreproducible subway token scheme that is not quite satisfactory because the validation process must be kept secret. Counterfeiting remains infeasible, however, given any number of valid subway tokens, even under unlimited computing power. Nonetheless, the scheme would be compromised should an enemy steal a validation turnstile.

The heart of these quantum subway tokens is an array of 20 pairs of mirrors, each pair containing a single trapped polarized photon with definite polarization direction chosen from among the four directions $0°$ (↔), $45°$ (↗), $90°$ (↕), and $135°$ (↖). The mirrors should be reflective enough to store the polarization information for a reasonable length of time. Such a sequence of trapped photons (or other two-state quantum systems) is called a VOM (verify-only memory) because it can be verified, but not accurately read or copied by someone ignorant of its contents. Each individual photon is called a vit.

To verify the sequence in the VOM, it suffices to measure each vit with a polarizer set to make the photon behave deterministically, for example reading the $0°$ and $90°$ photons with a $0°$ polarizer and the $45°$ and $135°$ photons with a $45°$ polarizer. The fact that all photons were absorbed or transmitted as expected would confirm the validity of the VOM, in the sense that a VOM differing in too many positions would have but a small probability of behaving in the same way.

A counterfeiter ignorant of the VOM's contents, on the other hand, could not avoid measuring at least some of the photons neither parallel nor perpendicular to their prepared polarizations, thereby causing the photons to behave probabilistically and losing their stored information.

Suppose the counterfeiter goes ahead anyway, making some measurement and preparing a new VOM whose photons agree with the result of the measurement. Then, for each photon, the counterfeiter has a 50% chance of making the wrong measurement and in this case there is a 50% chance that the incorrectly forged photon will give the wrong answer when subjected to subsequent attempted verification. Thus the entire counterfeit VOM has only $(3/4)^{20}$, or about 0.3% chance of passing inspection.

Besides its VOM, a subway token needs to contain an ordinary machine-readable data string to enable the turnstile to know which quantum measurements to make. This data could be a unique serial number enabling the turnstile to look up the expected VOM contents in a master list stored in each turnstile. More elegantly, the VOM contents could correspond to a computationally secure authentication tag,[5] which is computed from the data string together with some secret information known only of the turnstiles and the transit authority. Notice that this solution no longer offers security against unlimited computing power.

In this scheme, as in the others to be proposed later, the tokens must be distinct. A counterfeiter with access to a large number (20 would suffice) of tokens, known beforehand to be identical, could break the scheme by making both sets of measurements, and thereby learn the true VOM contents with only a small probability of error. Once this were known, the counterfeiter could make arbitrarily many copies of the given token.

Table 1. The polarization direction of photons in a quantum memory.

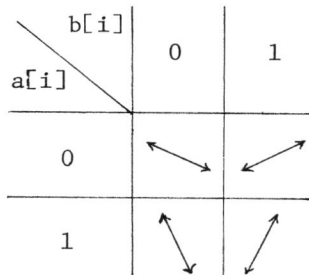

QUANTUM MULTIPLEXING

In order to obtain subway tokens that cannot be forged even if the validation mechanism is known, more sophistication is required. The major novel idea in this paper is that of Quantum Multiplexing. A quantum memory is a device capable of holding two pieces of information in such a way that either one can be read easily, but it should be computationally and/or physically infeasible to recover both. In particular, such a memory cannot be duplicated.

In order to multiplex messages A and B, we first expand them into A′ and B′, using an error correcting code that allows A and B to be recovered even if 14.7% of the bits are wrong. Let $a[i]$ and $b[i]$ denote the i-th bit of A′ and B′ respectively. For each i, the pair of bits $(a[i], b[i])$ is encoded by one single photon whose polarization angle is $(7-6b[i]-2a[i]+4a[i]b[i]) \times 22\frac{1}{2}$ degrees. In other words, the angle of the photon is given in Table 1.

If one tests each photon's vertical polarization, each bit of A′ will be recovered with an error probability of $\sin^2 22\frac{1}{2}°$, which is less than 14.7%. The message A can then be reconstructed, thanks to the error correcting code. Similarly, message B can be recovered if one tests the quantum memory's diagonal polarization.

For instance, if A and B are encoded into A′ = 1 0 1 1 0 1 0 1 and B′ = 0 0 1 0 1 0 1 1, then the photons will be polarized as follows: ↙ ↘ ↗ ↙ ↖ ↗ ↖ ↗ . A vertically polarized filter may read them as follows: 1 0 1 0 0 1 0 1, with one error on the fourth bit, which will be of no consequence since the error correcting code will allow us to recover the original message A.

Of course, in order to have a reasonably small probability of failure with the error correcting code, it will be necessary to encode long enough messages. To be safe, we should also use an error correcting code capable of recovering messages with somewhat more than 14.7% of the bits wrong. Care must be taken, however, for too much redundancy might allow both messages to be recovered.

Any attempts to cheat by testing intermediate polarization angles would only succeed in loosing both messages irreversibly, as long as each photon is measured independently of the others. In principle, however, there exist very complicated measurements that allow recovery of both messages by causing all the photons to interact simultaneously and coherently with the measuring apparatus. Although possible in principle, such measurements would be completely beyond the reach of present-day technology. We are currently investigating the hypothesis that this would indeed require a measuring apparatus of design computational complexity or physical bulk exponential in the length of the multiplexed messages. More details on this threat will appear in the final version of the paper.

UNFORGEABLE SUBWAY TOKENS

We are now ready to describe the unforgeable subway token scheme. Once and for all, the Transit Authority Administrator randomly selects two distinct large prime numbers congruent to 3 modulo 4, and computes their product. The latter, call it n, is revealed to the validating turnstiles. As we shall see, it will be sufficient to know n in order to validate tokens, yet knowledge of its factorization will be required to create them.

A unit is a triple <x,y,a> such that $0 < x < n/2$, $0 < y < n/2$, $0 < a < n$, a is relatively prime to n, the Jacobi symbol of x is plus one, and the Jacobi symbol of y is minus one. A unit is valid if $x^2 \equiv y^2 \equiv a \pmod{n}$. It is half valid if either x or y is a square root of a modulo n. Number theory tells us that valid units are plentiful as exactly one quarter of all numbers relatively prime to n have two distinct square roots modulo n that are below n/2, and these roots have complementary Jacobi symbols.[6] Moreover, it was shown by Rabin[7] that it is easy to come up with valid units as long as the prime factorization of n is known, whereas knowledge of a single valid unit gives away the factorization of n. On the other hand, knowledge of n is sufficient to create half valid units. Under the assumption that it is infeasible for the enemy to discover the factors of n, it is therefore clear that none but the transit authority can compute valid units.

Whereas units are mathematical concepts, elements are their physical quantum implementation. An element consists of a classical memory, together with a quantum multiplexing memory. The classical memory records the field a of some unit. The quantum memory multiplexes the fields x and y of the same unit. An element is (half) valid of such is the case with its underlying unit. The validation process of an element goes as follows. A random decision of reading either x or y from the quantum memory is made, its Jacobi symbol is verified, and its square modulo n is computed and checked against a. The validation process succeeds if no errors are found. It should be obvious that the validation process always succeeds on valid elements, whereas it succeeds with a 50% probability on half valid elements.

We have already seen that an enemy cannot compute units (hence create elements) that are better than half valid, short of factoring n. The key observation is that, thanks to the unique features of quantum multiplexing, this remains true even given unlimited supplies of distinct valid elements. Indeed, the only information obtainable from a valid element is a pair of numbers such that one is the square of the other (modulo n). But, of course, such pairs can easily be computed without reading valid elements. In other words, elements can be validated with a 50% chance of being cheated, but they can neither be created nor reproduced.

In order to reduce the probability of being cheated, a subway token consists of a collection of twenty valid elements. In order to validate a token, the turnstile randomly chooses, independently for each element on the token, which half of this element's quantum memory should be read for validation. The best a forger could produce under such circumstances is a token composed of twenty half valid elements. The turnstile would therefore decide to read precisely the valid entries, hence accept the forged token, with a probability smaller than one millionth. This should discourage the most daring forgers. It is also possible for the forger, as we leave the reader find by himself, to convert a deterministically sure $19 into a probabilistic value of $10.

Finally, we would like to point out a free bonus gained from the utilization of this unforgeable subway token scheme. Should a would-be forger steal a turnstile in the hope of forging tokens, we have seen that he would not get any useful information from his felony. It is amusing to realize that his efforts would have been a complete waste since it will not even be possible for him to reuse the already validated tokens found inside the turnstile: the mere fact that these tokens have been validated by the stolen turnstile implies that their relevant information has been already destroyed!

CONVENTIONAL IMPLEMENTATIONS OF VOM AND MULTIPLEXING

The effects of quantum multiplexing can be achieved to a large extent through the use of more conventional devices, such as shielded, tamper proof, shift registers. Let x and y be two length n messages to be multiplexed in the quantum memory sense. Consider a $5n$ bit long shift register such that only the n middle bits can be read from the outside. The register is initialized with x as the n leftmost bits, y as the n rightmost bits, and zeros in the $3n$ middle bits. Bitwise left and right shifts can be requested from the outside. Clearly, it is easy to gain access to either x or y: shift the register $2n$ bits to the right (for x) or to the left (for y) and look through the middle bit window. Moreover, when the first bit of x appears in the window, the last bit of y has already been irreversibly lost through the right end of the shift register.

Although more practical, this implementation could perhaps be fooled. For one thing, how could one ever be 100% sure that the shift register is indeed tamper proof? Perhaps a new kind of ray could violate its contents without the register sensing it. Another potential loophole in this implementation is that some measurable phenomenon could leak out when bits "fall off" the shift register.

Rather curiously, it seems that a conventional implementation of the simpler VOM is somewhat more complicated than a mere shift register. We leave to the reader the problem of finding how to do it.

CONCLUSION

The more conventional VOM and multiplexing memories discussed above would allow practical implementation of unforgeable subway tokens. Their unforgeability would however be based on current technological limitations. Similarly, David Chaum has proposed a fairly different, more economical, solution to the technologically unforgeable subway token problem.[8] On the other hand, the quantum subway tokens proposed in this paper offer protection against technological breakthroughs, but they could not be built with today's technology. The best available device known of the authors for holding quantum information is capable of preserving it for just over a second. However, the continuing advance of cryogenic and optical techniques promises considerably longer life time in the future.

REFERENCES

1. W. Diffie and M.E. Hellman, New Directions in Cryptography, IEEE Trans. Info. Th., IT-22:644 (1976).
2. R.L. Rivest, A. Shamir and L. Adleman, On Digital Signatures and Public-Key Cryptosystems, CACM, 21: 120 (1978).
3. G.B. Kolata, New Codes Coming into Use, Science Magazine, 208:694 (1980).
4. P.A.M. Dirac, "The Principles of Quantum Mechanics, 4th edition," Oxford University Press (1958).
5. G. Brassard, Computationally Secure Authentication Tags Requiring Short Secret Shared Keys, in: "Advances in Cryptography: Proceedings of CRYPTO 82," R. Rivest, ed., Plenum Press, New York (1983).
6. M. Blum, Coin Flipping by telephone: A Protocol for Solving Impossible Problems, in: "Proceedings of 24th Compcon," IEEE, New York (1982).
7. M.O. Rabin, Digitalized Signatures and Public-Key Functions as Intractable as Factorization, MIT/LCS/TR-212 (1979).
8. D. Chaum, personal communication (1982).

Session V: Special Session on Cryptanalysis

A POLYNOMIAL TIME ALGORITHM FOR BREAKING THE BASIC MERKLE-HELLMAN CRYPTOSYSTEM

(Extended abstract)

Adi Shamir

Applied Mathematics
The Weizmann Institute
Rehovot, Israel

ABSTRACT

The cryptographic security of the Merkle-Hellman system (which is one of the two public-key cryptosystems proposed so far) has been a major open problem since 1976. In this paper we show that when the elements of the public key $a_1, \ldots, a_n$ are modular multiples of a superincreasing sequence (as proposed by Merkle and Hellman), almost all the equations of the form

$$\sum_{i=1}^{n} x_i a_i = b \qquad x_i \in \{0,1\}$$

can be solved in polynomia time, and thus the cleartexts $x_1 \ldots x_n$ that correspond to given ciphertexts b can be easily found.

OUTLINE OF THE ALGORITHM

The algorithm proposed in this paper analyses the given numbers $a_1,\ldots,a_n$ and attempts to find a <u>trapdoor pair</u> of natural numbers W and M such that Wa_i (mod M) is a superincreasing sequence and its sum is smaller than M. Knowledge of any pair of numbers with these properties makes it possible to solve arbitrary equations of the form

$$\sum_{i=1}^{n} x_i a_i = b \qquad x_i \in \{0,1\}$$

in polynomial time (see Merkle and Hellman [1978]). Since the a_i were obtained from a superincreasing sequence by modular multiplication, we know that at least one such pair exists. Our algorithm finds some trapdoor pair, but it is not guaranteed to find the original pair used in the construction of the a_i's.

In the Merkle-Hellman construction, the elements of the original superincreasing sequence have known sizes (but unknown values!). For the sake of simplicity, we assume that the i-th number has $n + i - 1$ bits, so that the smallest element is smaller than 2^n, the largest element is smaller than 2^{2n-1}, and the modulus is between 2^{2n-1} and 2^{2n} (in their original paper, Merkle and Hellman recommend this scheme with $n = 100$). After the modular multiplications, all the numbers become approximately 2n-bit long. They can be published in a permuted order (so that a_1 does not necessarily correspond to the smallest element in the superincreasing sequence), but our algorithm remains polynomial in n even when such an unknown permutation is used.

The algorithm is divided into two parts. In the first part, Lenstra's integer programming algorithm is used to find a rational number $0 < \alpha < 1$ such that a <u>necessary</u> condition for W and M to be a trapdoor pair is that $\frac{W}{M} \in [\alpha, \alpha+\varepsilon]$ for a certain small ε. In the second part, we use the fact that the ratio $\frac{W}{M}$ is approximately known to find at most n^2 subintervals (ℓ_i, r_i) in $[\alpha, \alpha+\varepsilon]$ such that $\frac{W}{M} \in (\ell_i, r_i)$ for some i is also a <u>sufficient</u> condition for W, M to be a trapdoor pair. If we assume that some pair exists, at least one of the subintervals must be non-empty. By using a fast diophantine approximation algorithm, we can find the smallest W and M whose ratio is in such a subinterval.

Let W_0, M_0 be the (unknown) trapdoor pair of 2n-bit numbers used in the construction of the a_i sequence. The first step of the algorithm is to generalize the definition of a trapdoor pair to arbitrary real positive W and M. When $M = M_0$, the graph of the function $W a_i \pmod{M_0}$ for real multipliers $0 \leq W < M_0$ has a sawtooth form:

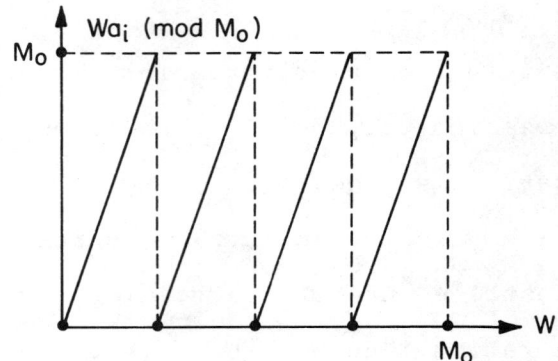

The slope of the function (except at discontinuity points) is a_i, the number of minima is a_i, and the distance between successive minima is M_0/a_i (which is slightly more than 1).

If a_i corresponds to the smallest element in the superincreasing sequence, then the multiplier W has the property that $Wa_i \pmod{M_0}$ is at most 2^n, and thus the distance between W and the closest minimum to its left cannot exceed $2^n/a_i \approx 2^{-n}$. The unknown W must thus be very close to some minimum of the sawtooth curve. Unfortunately, even if we impose the integrality constraint on W (which we do not), there are too many possible values for W and we cannot check them one by one.

If a_j corresponds to the second smallest element in the superincreasing sequence, then a similar analysis shows that W must also be within a distance of $2^{n+1}/a_j$

from a minimum of the a_j-sawtooth, and thus the two minima of a_i and a_j must be very close to each other. This greatly reduces the number of places in which W may be, but it still does not characterize it uniquely.

We can proceed in a similar way and superimpose more sawtooth curves on the same diagram. The fact that W is close to a minimum on each curve implies that all these minima are close to each other, and thus we can replace the problem of finding W by the equivalent problem of finding the accumulation points of minima of the various curves.

In the full paper we show that when four sawtooth curves are superimposed, the probability that four minima will be so close to each other is so small that it is extremely unlikely to happen in more than a few places in the region $0 \leq W < M_0$ (it must happen somewhere since by the construction of the a_i's such a W_0 exists). The number 4 is independent of n, and depends only on the ratio between the sizes of M_0 and the smallest element of the superincreasing sequence (which was assumed to be 2).

Two problems remain: How to get rid of M_0 (whose value is actually unknown) and how to find the accumulation point of the minima of the four sawtooth curves.

The key observation is that the location of the accumulation point in the diagram depends on the <u>slopes</u> of

the curves, but not on their sizes. If we divide both
coordinates by M_0, we get the sawtooth curve of the
function Va_i (mod 1), $0 \leq V < 1$, which is independent
of M_0:

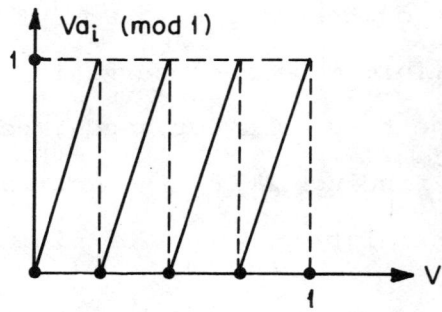

In the new coordinate system the slope of the curve remains
a_i, the number of minima remains a_i, but the distance
between successive minima is reduced to $1/a_i$. The original W parameter is replaced by a new $V = W/M_0$ parameter, and the allowable distance between this parameter
and the closest curve minimum is reduced by a factor of
approximately 2^{2n} (from 2^{-n} to 2^{-3n}).

The problem of locating the accumulation point of
minima in the new coordinate system can be described by
linear inequalities with four integral unknowns. Without loss of generality, we assume that a_1 a_2 a_3 a_4 correspond to the four smallest elements in the superincreasing sequence (there are $O(n^4)$ ways to guess them). We
further assume that among the four minima at the accumula-

tion point, the a_1-minimum is the rightmost (i.e., closest to W_0/M_0). Then the conditions that the i-th minimum of a_1, j-th minimum of a_2, k-th minimum of a_3, and ℓ-th minimum of a_4 are sufficiently close to each other are:

$$i, j, k, \ell \text{ integers} \qquad 1 \leq i \leq a_1 - 1$$

$$0 \leq \frac{i}{a_1} - \frac{j}{a_2} \leq 2^{-3n+1} \qquad 1 \leq j \leq a_2 - 1$$

$$0 \leq \frac{i}{a_1} - \frac{k}{a_3} \leq 2^{-3n+2} \qquad 1 \leq k \leq a_3 - 1$$

$$0 \leq \frac{i}{a_1} - \frac{\ell}{a_4} \leq 2^{-3n+3} \qquad 1 \leq \ell \leq a_4 - 1$$

By multiplying the inequalities by their denominators, we get an equivalent system in which all the coefficients of i, j, k and ℓ are integers with no more than 5n bits. Since Lenstra's integer programming algorithm is polynomial in the size of the coefficients for a fixed number of unknowns, we can find the (almost certainly unique) accumulation point of the four minima in polynomial time.

Once the value of i is known, it is easy to find the interval $[\alpha, \alpha+\varepsilon]$ of V values for which the values of all the n sawtooth curves are properly bounded. An important property of this interval is that it cannot contain discontinuity points since all the sawtooth values in it must be smaller than 1.

A typical enlarged section of the superimposed

diagram in the vicinity of W_0/M_0 is:

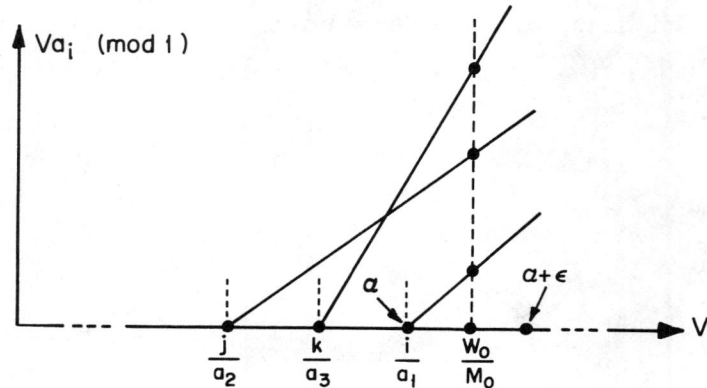

Any pair of numbers W, M such that $\frac{W}{M} \in [\alpha, \alpha+\varepsilon]$ gives properly bounded values under modular multiplication, but these values need not be a superincreasing sequence and thus they do not necessarily lead to an easily solvable knapsack. The second part of the algorithm extracts from the $[\alpha, \alpha+\varepsilon]$ interval those subintervals (ℓ_i, r_i) for which the transformed sequence is guaranteed to be superincreasing.

Since the $[\alpha, \alpha+\varepsilon]$ interval does not contain discontinuity points, the n sawtooth curves look like n linear segments in it. These n segments can intersect each other in at most $O(n^2)$ points. By finding and sorting these points, we can subdivide $[\alpha, \alpha+\varepsilon]$ into $O(n^2)$ subintervals with a well defined vertical order between the curves in each subinterval. When this order is known, we can express the conditions for a superincreasing sequence by the linear inequalities

$$\left(Va_{\pi(i)} - c_{\pi(i)}\right) > \sum_{\pi(j)<\pi(i)} \left(Va_{\pi(j)} - c_{\pi(j)}\right)$$

$$\sum_{i=1}^{n} (Va_i - c_i) < 1$$

in which the π is the permutation of the indices specified by the vertical ordering in the subinterval and the c_i is the number of a_i-minima between 0 and the accumulation point. The solution of each set of inequalities is a (possibly empty) subinterval (ℓ_i, r_i) in which all the superincreasing and size conditions are satisfied. At least one of these subintervals must be non-empty, and the smallest natural numbers W and M such that W/M belongs to such an interval can be found in polynomial time (note that W and M cannot exceed W_0 and M_0, which are 2n-bit long). Once these numbers are found, the cryptanalysis of arbitrary ciphertexts in the $a_1,\ldots,a_n$ system becomes trivial.

CONCLUSIONS AND OPEN PROBLEMS

We have demonstrated that Merkle-Hellman cryptosystems in which the public keys are obtained from superincreasing sequences by a single modular multiplication are totally insecure. It remains an open problem whether keys obtained by two or more modular multiplications are cryptographically secure. In addition, the exact complexity of our algorithm needs further analysis since it can be optimized in a number of ways (e.g., 3 superimposed sawtooth curves are sometimes enough, and then the Lenstra algorithm can be replaced by a much simpler algorithm based on continued fractions).

BIBLIOGRAPHY

Merkle, R., and Hellman M.,[1978],"Hiding information and signature in trapdoor knapsacks," IEEE Trans. Information Theory, IT-24-5, September 1978.

A PRELIMINARY REPORT ON THE CRYPTANALYSIS OF
MERKLE-HELLMAN KNAPSACK CRYPTOSYSTEMS[†]

E. F. Brickell, J. A. Davis and G. J. Simmons

Sandia National Laboratories
Albuquerque, New Mexico 87185

INTRODUCTION

In April, 1982, Adi Shamir caused a furor with the announcement [1] of "A Polynomial Time Algorithm for Breaking Merkle-Hellman Cryptosystems." Like many others who received his "extended abstract," members of the mathematics department at the Sandia National Laboratories undertook a careful study of both the algorithm and the underlying mathematical concepts. This paper summarizes some of our findings. In order to meet the deadline for Crypto'82 the style will be deliberately telegraphic -- and informal. A complete paper will be presented this fall at the Twelfth Conference on Numerical Mathematics and Computing at Winnipeg, the proceedings of which will be published in Congressus Numerantium. It should also be remarked that in discussions with Adi at Crypto'82, we learned that some of the results presented here as new were also arrived at independently by him in the period since his original announcement.

Finally, we shall adopt Adi's notation, etc., wherever possible, so as to assist the reader in making the transition from one paper

---

[†] This work performed at Sandia National Laboratories supported by the U. S. Department of Energy under contract number DE-AC04-76DP00789.

to the other. Since Adi's paper was given in the same session at Crypto'82 as this one, and will accompany it in the Proceedings, we assume the reader to be familiar with Adi's development and notation, and will only discuss those points that are essential to illustrate arguments made here.

A WAY OUT FOR THE KNAPSACK DESIGNER

In this section we show how the designer of a Merkle-Hellman knapsack can, with a feasible amount of precomputation, generate a hard knapsack that is immune to the cryptanalytic algorithm described in Adi's "extended abstract."

Given an n weight hard knapsack $\{a_i\}$ derived from a Merkle-Hellman easy knapsack [2] (super increasing sequence of weights) by the modular (trapdoor) transformation:

$$a_i \equiv W_0 b_i \quad (\mod M_0) \ .$$

Adi normalizes the inverse modular functions

$$f_i(z) = a_i z - \left[\frac{a_i z}{M_0}\right] M_0$$

by dividing out the unknown (to the cryptanalyst) modulus M_0 to obtain the functions

$$f_i(v) = a_i v - [a_i v] = a_i v - c_i$$

where $0 \leq v = \frac{z}{M_0} \leq 1$, $0 \leq f_i(v) \leq 1$ and c_i is an integer, $c_i < a_i$. The $f_i(v)$ are piecewise linear (sawtooth) functions with slope a_i having a_i zeroes uniformly spaced $\frac{1}{a_i}$ apart on the v axis. At $v_0 = \frac{W_0^{-1}}{M_0}$, i.e., at the multiplicative inverse of W_0, one of the functions must have a value $< \frac{2^n}{M_0} \approx 2^{-n}$, since $b_1 < 2^n$ in a Merkle-Hellman easy knapsack. Hence the minima of $f_1(v)$ closest to v_0 on the left cannot be further away from v_0 than $\frac{2^n}{M_0 a_1} \approx 2^{-3n}$. Note that we have assumed a_1 is the modular image of b_1 in accordance with Adi's usage. Similarly the closest minima to the left of v_0 of $f_i(v)$, i = 2, 3 and 4 (corresponding to b_2, b_3 and b_4) are no further away than $\frac{2^{-n+i-1}}{M_0 a_i}$. The value of v_0 is of course unknown, and in fact is what the cryptanalyst is trying to localize, however the distance between pairs of the minima of the $f_i(v)$, $i = 1, 2, 3$, and 4, closest to v_0 can be expressed. Adi uses approximations of these distances to form three inequalities.

$$0 \leq \frac{c_1}{a_1} - \frac{c_2}{a_2} \leq 2^{-3n+1}$$
$$0 \leq \frac{c_1}{a_1} - \frac{c_3}{a_3} \leq 2^{-3n+2} \qquad (1)$$
$$0 \leq \frac{c_1}{a_1} - \frac{c_4}{a_4} \leq 2^{-3n+3}$$

where $0 \leq c_i$ is the number of minima of $f_i(v)$ between 0 and a value of v where the four functions simultaneously satisfy $f_i(v) < 2^{-n+i-1}$. The algorithm is based on the expectation that if for some choice of four weights from $\{a_i\}$ the minima of the $f_i(v)$ do accumulate so close together, i.e., if four integers c_i exist satisfying (1) then this "(almost certainly unique)" point is close to $\frac{W_0^{-1}}{M_0}$.

In keeping with the stated objective of this short paper, we ignore questions of uniqueness, etc., and concentrate instead on how the knapsack designer can cause the system (1) to fail to have a solution. Consider one of the functions $f_i(v)$ in the vicinity of v_0.

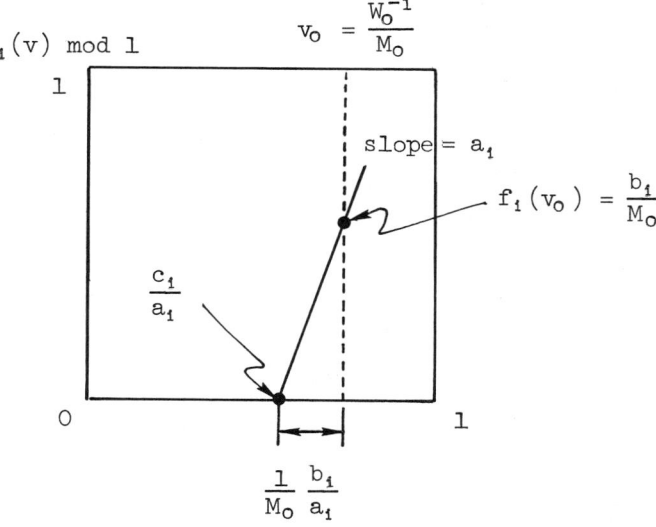

Since v_0 is the normalized multiplicative inverse of W_0, $f_i(v_0) = \frac{b_i}{M_0}$ for all i. The slope of $f_i(v)$ is a_i everywhere except at the zeroes of $f_i(v)$, hence at v_0 in particular. The distance between v_0 and the minima of $f_i(v)$ closest to v_0 on the left is therefore precisely

$$\frac{b_i}{a_i} \frac{1}{M_0} = v_0 - \frac{c_i}{a_i}$$

so that the closest minima of $f_i(v)$ are located at

$$\frac{c_i}{a_i} = v_0 - \frac{b_i}{a_i} \frac{1}{M_0} \quad .$$

If these values are substituted into system (1) and all terms multiplied through by M_0, one obtains the equivalent system of inequalities

$$0 \leq \frac{b_2}{a_2} - \frac{b_1}{a_1} \leq M_0 2^{-3n+1}$$

$$0 \leq \frac{b_3}{a_3} - \frac{b_1}{a_1} \leq M_0 2^{-3n+2} \quad . \quad (1')$$

$$0 \leq \frac{b_4}{a_4} - \frac{b_1}{a_1} \leq M_0 2^{-3n+3}$$

The essential point on which Adi's algorithm hinges is that the expected values of the ratios $\frac{b_i}{a_i}$, for the smallest weights in a Merkle-Hellman knapsack, are extremely small; i.e., $\frac{b_1}{a_1} \approx 2^{-n}$, $\frac{b_2}{a_2} \approx 2^{-n+1}$, etc. Consequently, with a very high probability, for randomly constructed trapdoor knapsacks, the difference of the smallest ratios are correspondingly small. The form of (1'), and the preceding remarks, however, suggest a way out for the knapsack designer. Roughly speaking, if he can cause one of the weights b_2, b_3, or b_4 to map into an unexpectedly small a_i, the slope of the corresponding $f_i(v)$ will be smaller than expected and the closest minima of $f_i(v)$ will be further to the left of v_0 than expected -- which can cause one or more of the inequalities in (1') to fail. We make this intuitive notion precise in two ways.

First, the designer can force one of the inequalities in (1') to fail by solving the linear inequalities

$$\left.\begin{array}{c} \dfrac{M_0}{2} < W_0 b_1 - c_1 M_0 = a_1 < M_0 \\ 0 < W_0 b_i - c_i M_0 = a_i < A \\ 2^{2n-1} < M_0 < 2^{2n} \\ W_0 < M_0 \end{array}\right\} \quad (2)$$

for one of the a_i = 2, 3 or 4 -- where A is chosen to cause $\dfrac{b_i}{a_i}$ to be larger than expected. The result -- as seen from (1') -- will be to cause the distance between the minima of $f_i(v)$ and $f_1(v)$ closest to v_0 to be too far apart for the corresponding inequality in (1) to be satisfied. Computationally this is an easy problem for the designer -- so long as he need only force a small number of values of the a_i (only one in the present discussion) to fall in a restricted range, i.e., < A. We next examine the likelihood that (2) has a solution.

Under the assumption that the values $\dfrac{a_i}{M_0}$ resulting from the modular mappings

$$a_i \equiv W_0 b_i \quad (\bmod\ M_0)$$

for randomly chosen W_0 and M_0, subject to the already stated constraints, are well approximated by independent uniformly distributed random variables, A_i, on the unit interval, we have proven the following theorem.

<u>Theorem 1</u>

For $b_j > b_i$

$$P_{i,j}(z) = P\left[\left(\dfrac{b_j}{A_j} - \dfrac{b_i}{A_i}\right) > z\right] = \dfrac{b_i}{zM_0}\left[\alpha - \dfrac{b_i}{zM_0}\ln\left(1 + \alpha\dfrac{zM_0}{b_i}\right)\right] + (1-\alpha) \quad (3)$$

where

$$\alpha = \begin{cases} 1 & \text{if}\ zM_0 \geq b_j - b_i \\ \dfrac{b_i}{b_j - zM_0} & \text{otherwise} \end{cases}$$

A Monte Carlo Evaluation of (3) using $z = M_0 2^{-3n+1}$ from (1')
and randomly drawn Merkle-Hellman easy weights b_i, modulii M_0 and
multipliers W_0 yielded the probabilities

$$P_{12}(z) = 0.30$$
$$P_{13}(z) = 0.36$$
$$P_{14}(z) = 0.41$$

for $n \geq 50$. In other words the probability that the system of
inequalities (1) would naturally fail to have a solution is $P_f \approx 0.74$.
Even more significant though is the ease with which the designer can
find a solution to system (2) which will result in a hard knapsack
that is immune to cryptanalysis by algorithm (1).

AN ALTERNATE APPROACH TO INVERTING MERKLE-HELLMAN KNAPSACKS

In the preceding section we showed that it is not necessarily
true that in the vicinity of a point at which the four functions
$f_i(v)$ are simultaneously very small that all of the minima of the
$f_i(v)$ are also very close together. It is true, however, that if
one considers some fraction αn of the $f_i(v)$ whose values all satisfy

$$f_i(v_0) < 2^{-n+\alpha n}$$

at v_0, that many, but not necessarily all, of the closest minima
will accumulate near v_0. We shall not pursue this approach further
-- except to remark that we have proven theorems that strongly suggest that finding k minima of the $f_i(v)$ crowded sufficiently close
together is a good indication of a neighborhood to a solution -- and
that the knapsack designer appears to be faced with a computationally infeasible task in eluding this algorithm with $k = 4$ or 5,
$\alpha \approx 1/6$ and $n \approx 100$.

Instead of looking for accumulations of minima of the $f_i(v)$ as
indicators of regions in which solutions have a high probability of
existing, we focus instead on the distribution of the values of

$f_i(v)$. At v_0, since the a_i are modular images of a super increasing sequence, the values of the $f_i(v_0)$ are exponentially dense near 0. On the other hand at values of v away from v_0 the values of the $f_i(v)$ are nearly uniformly distributed in the unit interval. What this says is that if we consider a bound $B \ll 1$, then in the vicinity of v_0, there will be $\approx n + \log_2 B$ of the $f_i(v)$ whose values are simultaneously less than B while away from v_0 the probability of $f_i(v) < B$ for any i will be roughly proportional to B. We shall make these remarks more precise in a moment -- but the basic idea should be clear. Choose B sufficiently large that the $\ell \approx n + \log_2 B$ of the $f_i(v)$ that have values less than B in the neighborhood of v_0 due to the exponential accumulation of function values at v_0 are a fixed fraction, αn, of the total, but small enough that the probability of k ($k = 4$ or 5) of those αn $f_i(v)$ being less than B at points away from v_0 is essentially zero.

It might at first seem an impossible task to search the continuous unit interval for values of v at which the $f_i(v)$ accumulate, however we can greatly simplify the task by the following argument that reduces the search to a finite -- albeit large -- number of values of v. Given k weights from the hard knapsack and a bound $B \ll 1$, assume that at some point v' all k of the related functions satisfy $f(v') < B$. The inequalities will continue to hold for all $v'' \le v \le v'$, where v'' is the location of the closest minima below v' of any of the $f_i(v)$ since at v'' at least one of the functions has value zero and the others have decreased in value by $a_i(v'-v'')$.

$$v'' = \max_{i=1}^{k} \left\{ \frac{[a_i v']}{a_i} \right\}.$$

The search for points at which the functions simultaneously have values less than B can therefore be restricted to those points at which one or more of the $f_i(v)$ have minima, i.e., rationals of the form

$$\frac{c_i}{a_i} \quad \text{where} \quad \begin{cases} 0 \le c_i < a_i \\ 1 \le i \le k \end{cases}.$$

As already remarked there are a_i such minima of $f_i(v)$, however the total number of minima for a collection of functions may be significantly less than the sum of the associated a_i since some points may be multiple minima.

Given a subset of k weights, A, define the functions S_j;

$$\left.\begin{array}{l} S_1 = \displaystyle\sum_{a_i \in A} (a_i) - \binom{k}{1} \\[6pt] S_2 = \displaystyle\sum_{\substack{a_i, a_j \in A \\ a_i \neq a_j}} (a_i, a_j) - \binom{k}{2} \\[6pt] \vdots \\[2pt] S_j = \displaystyle\sum (\text{GCD of } j \text{ element subsets of } A) - \binom{k}{j} \\[6pt] \vdots \\[2pt] S_k = (A) - \binom{k}{k} \end{array}\right\} \quad (4)$$

If $\hat{a} = \max\{a_i\}$ then trivially $f_i(v) < B$ in the interval

$$\left(0 < v < \frac{B}{\hat{a}}\right).$$

Rationals in this interval are solutions for the hard knapsack only in the (improbable) event that the $\{a_i\}$ are already a super increasing sequence, i.e., the "hard" knapsack is in fact trivial. We therefore exclude the open interval $(0 < v < \frac{1}{\hat{a}})$ at the origin in the following discussion.

Theorem 2

Given a k element subset A of the weights $\{a_i\}$, the number of points at which precisely j of the associated functions $f_i(v)$ have minima is given by

$$N_j = \sum_{i=0}^{k-j}(-1)^i \binom{j+i}{j} S_{j+1} \qquad 1 \leq j < k$$

and the total number of minima by (5)

$$N_{min} = \sum_{i=1}^{k} N_i = \sum_{i=1}^{k}(-1)^{i-1} S_i$$

For a randomly drawn set of k weights, the expected number of open intervals in which all k of the associated functions have values less than B is therefore

$$N_s = \sum_{i=1}^{k} N_i \cdot 2^{(k-i)\log_2 B} \qquad (6)$$

on the assumption that those functions not equal to zero are uniformly -- and independently -- distributed on the unit interval. The detailed behavior of the N_i is an exceedingly difficult number theoretic question -- hinging as it does on the distribution of common divisors among the elements of $\{a_i\}$, however, the following argument shows that k is greater than 3 and is almost certainly no greater than five under our assumptions about the a_i.

For the purposes of illustration, let $B = 2^{-n+n/6}$, so that $\approx n/6$ of the $f_i(v_0)$ have values less than $2^{-5/6\,n}$ at a point away from v_0 and 0 is

$$P_f = \frac{N_s}{N_{min}} = \frac{1}{N_{min}} \sum_{i=1}^{k} N_i \, 2^{-5/6(k-i)n} \qquad (7)$$

If $N_1 \gg N_2 2^{-1 \log_2 B}$ so that the points at which only one function has a minima exponentially dominate the others, then N_s can be approximated by

$$N_s \approx 2^{(k-1)\log_2 B} \sum_{i=1}^{k} a_i$$

or by

$$N_s \approx 2^{n/6\,(17-5k)+\log_2 k}$$

in the present example. $k = 3$ and $n \geq 100$ results in an intolerable number of false alarms, while $k = 4$ -- if the assumption about N_1 exponentially dominating the N_j -- results in

$$N_s \approx 2^{-n/2+2} = 3.6 \times 10^{-15}$$

for the case of $n = 100$, i.e., an exceedingly small number of false solution intervals. Our experience with Monte Carlo computations bears out this expectation for randomly generated Merkle-Hellman knapsacks.

To summarize, what we have shown is that if the bound is taken to be $B = 2^{-5/6\,n}$, so that $n/6$ of the $f_1(v_0)$ are less than B, then the probability that even four randomly chosen $f_1(v)$ will be less than B at a point away from v_0 is very small.

It is worth remarking that choosing B so as to cause αn of the $f_1(v_0)$ to be less than B avoids the $O(n^4)$ search problem of Adi's original proposal. Since the method described here will be satisfied by any subset of k out of the αn a_i corresponding to the $f_1(v_0) < B$, the probability of hitting such a subset at random is

$$P_h = \frac{\binom{\alpha n}{k}}{\binom{n}{k}} = \alpha^k \prod_{i=1}^{k-1} \frac{(n-i\alpha^{-1})}{(n-i)} \,. \tag{8}$$

If $k = 4$ and $\alpha = 1/6$, then for all $n \geq 52$ the probability of hitting a subset of four of the desired a_i is

$$P_h \geq \frac{1}{2 \cdot 6^4}$$

or conversely the expected number of sets of four weights that need to be tried before a success is less than 2592. If $k = 4$ and $\alpha = 1/10$, then for all $n \geq 91$ the upper bound on the expected number of tries increases to 20,000.

With these preliminaries out of the way, we are now ready to describe the algorithm by which we propose to localize v_o. Given a set of four a_i, it has already been shown that it is only necessary to consider those points $\frac{c_i}{a_i}$, $0 \le c_i < a_i$, at which a minima of at least one of the $f_i(v)$ occurs. The values of the other functions at these minima are

$$f_j\left(\frac{c_i}{a_i}\right) = \left(\frac{c_i}{a_i}\right) a_j - \left[\frac{c_i}{a_i} a_j\right] = \frac{a_j}{a_i} c_i - c_j \quad . \tag{9}$$

Hence, the inequalities that need to be satisfied in order for all four functions to have values less than the bound B are:

$$\left.\begin{array}{c} 0 \le \dfrac{c_i}{a_i} - \dfrac{c_j}{a_j} \le \dfrac{B}{a_j} \\[6pt] 0 \le \dfrac{c_i}{a_i} - \dfrac{c_k}{a_k} \le \dfrac{B}{a_k} \\[6pt] 0 \le \dfrac{c_i}{a_i} - \dfrac{c_\ell}{a_\ell} \le \dfrac{B}{a_\ell} \end{array}\right\} \quad . \tag{10}$$

As we have already remarked, inequalities (10) are very similar in appearance to Adi's system (1). In fact, if we replace B by the tightest possible upper bound for the functions $f_i(v_o)$, $i = 2, 3$ and 4 and assume that the corresponding a_i are as large as possible we get precisely system (1). The important point is not that under these extreme conditions (10) is equivalent to (1), but rather that (10) describes a search (on the vertical axis) for accumulations of values of the $f_i(v)$ rather than a search (on the horizontal axis) for accumulations of minima of the $f_i(v)$.

We use the same substitution of values for the $\frac{c_i}{a_i}$ used earlier in obtaining the system (1') from (1) to obtain a system of inequalities equivalent to (10):

$$\left.\begin{array}{c} 0 \le \dfrac{b_j}{a_j} - \dfrac{b_i}{a_i} \le \dfrac{M_o}{a_j} B \\[6pt] 0 \le \dfrac{b_k}{a_k} - \dfrac{b_i}{a_i} \le \dfrac{M_o}{a_k} B \\[6pt] 0 \le \dfrac{b_\ell}{a_\ell} - \dfrac{b_i}{a_i} \le \dfrac{M_o}{a_\ell} B \end{array}\right\} \quad . \tag{10'}$$

When put in this form (10'), it is easy to see why systems (1') and (10') behave differently. In (10'), if an a_j, a_k or a_l is caused to be smaller than expected, which would have caused the corresponding inequality in (1') to fail to have a solution, the right-hand bound grows proportionally since the corresponding weight a appears in the denominator.

The weight a_i -- whose associated function $f_i(v)$ was assumed to have a minima at the test points -- is distinguished in (10), so that for each choice of four weights, there are four systems of the form of (10) whose solution must be attempted: letting each weight take the role of a_i in turn. Setting $B = 2^{-n+an}$ and multiplying the inequalities in (10) through by $2^n a_i a_j$, etc., we obtain the equivalent system --

$$\left. \begin{array}{l} 0 \leq 2^n a_j c_i - 2^n a_i c_j \leq a_i 2^{an} \\ 0 \leq 2^n a_k c_i - 2^n a_i c_k \leq a_i 2^{an} \\ 0 \leq 2^n a\, c_i - 2^n a_i c \leq a_i 2^{an} \end{array} \right\} \qquad (10')$$

in which the coefficients are integers with no more than 5n bits -- which can be solved in precisely the same way and with the same computational difficulty as in Adi's proposal.

All other aspects of this algorithm are the same as in Adi's method -- only the means of localizing the regions in which to test for solutions has been changed. The reader is referred to Adi's paper for the essential details involved in subdividing intervals and testing for super increasing sequences that satisfy the unique invertibility condition.

CONCLUSION

The reason that we have titled this paper "A Preliminary Report on the Cryptanalysis of Merkle-Hellman Knapsack Cryptosystems," is primarily our concern that just as the knapsack designer could elude cryptanalysis by Adi's original algorithm, that he may also be able to elude the alternative algorithm described here. Our investigation of this question has only just begun. However, there are several worrisome areas. For example, in the analysis of N_s and P_f, we used the assumption that the a_i were randomly distributed. The designer may be able to force the values of the a_i in such a way as to invalidate this assumption and hence to invalidate statements we have made about k, P_f, N_s, N_{min}, N_j, etc., that were true for randomly constructed $\{a_i\}$. Similarly, in (10) and (10'), there is an implicit assumption -- in the discussion of the solution to these systems which we bypassed by referring the reader to Adi's paper -- that not many of the a_i's corresponding to the αn, $f_i(v_0) < B$ are much smaller than expected -- in the as yet unproven expectation that it will be difficult for the designer to cause this assumption to fail and to at the same time make the hard knapsack still be "hard" against direct solution.

Finally, we conclude with the observation (caution) that it is extremely dangerous to impute properties to a deliberately designed cryptosystem -- even though these properties can be shown to hold with very high probability for almost all randomly chosen cryptosystems based on the same algorithm.

REFERENCES

1. Shamir, Adi, "A Polynomial Time Algorithm for Breaking Merkle-Hellman Cryptosystems," (extended abstract) 7 p. research announcement, April, 1982.

2. Merkle, R. and M. Hellman, "Hiding Information and Receipts in Trapdoor Knapsacks," IEEE Transactions on Information Theory IT-24 (1978), p. 525-530.

ON BREAKING THE ITERATED MERKLE-HELLMAN
PUBLIC-KEY CRYPTOSYSTEM

Leonard M. Adleman*

University of Southern California and
Massachusetts Institute of Technology

I. Introduction

In 1976 Diffie and Hellman introduced the concept of a public-key cryptosystem [1]. In 1977 Rivest, Shamir, and Adleman discovered the first incarnation of such a system [4], and soon afterwards Merkle and Hellman produced a second one [3]. Despite the widespread interest in the area, the years have produced no other public-key cryptosystems which have attracted widespread interest.

The Merkle-Hellman system is based on the knapsack problem, and in the original paper on the topic, both a basic method and an iterated method were presented. The iterated method was introduced "for improving the security and utility of the basic method." In April of 1982, Adi Shamir demonstrated that the basic knapsack cryptosystem was insecure [5]. In addition, Shamir states that the most important remaining open is the cryptographic security of the iterated systems. In this paper, we build upon Shamir's results to establish the insecurity of the iterated systems as well.

Our method of attack uses recent results of Lenstra and Lovacz [2]. We treat the cryptographic problem as a lattice problem, rather than a linear programming problem as in Shamir's result. Like Shamir, we are unable to present a rigorous proof that the algorithm works. However, an analysis of the algorithm in the presence of "reasonable

---
*Research sponsored by National Science Foundation, Grant #MCS-8022533

assumptions" will be presented. We will be particularly
concerned with defining a broad class of iterated systems
where the algorithm is virtually guaranteed to work. Our
Analysis involves consideration of the Lovacz-Lenstra
algorithm for finding "almost" the smallest vector in a
latice.

In addition we will apply a varient of the algorithm
to the Graham-Shamir public-key cryptosystem -- a variant
of the Merkle Hellman system. Finally, since the algorithm
has actually been implmented, examples of is performance
will be given.

II. Iterated Knapsack Systems

Public-key cryptosystems require the generation of a
"mated pair" of keys. One key is kept secret, the other
is made public. It is crucial that the problem of
computing the secret key from the public key be intractable.
In the iterated knapsack systems this is apparently not the
case. Below is a description of the procedure used to
generate a mated pair of keys for such a system. How these
keys are used for encryption and decryption will not
concern us.

Step 0

Generate a sequence of natural numbers $a_{0,1}, a_{0,2}, \ldots, a_{0,n}$ such that

$$a_{0,i} \geq \sum_{j=1}^{i-1} a_{0,j} \qquad i = 2, 3, \ldots, n$$

(such a sequence is said to be "super-increasing").

Step 1

Generate numbers W_1, M_1 such that

a) $M_1 \geq \sum_{i=1}^{n} a_{0,i}$

b) $(W_1, M_1) = 1$.

Compute

$$a_{1,i} \equiv W_1 a_{0,i} \text{MOD}(M_1)$$

.
.
.

Step z

Generate numbers W_z, M_z such that

a) $M_z \geq \sum_{i=1}^{n} a_{z-1,i}$

b) $(W_z, M_z) = 1$

Compute

$$a_{z,i} \equiv W_z a_{z-1,i} \text{MOD}(W_z)$$

The Secret Key is

$$<<a_{0,1}, \ldots, a_{0,n}>, M_1, W_1, M_2, W_2, \ldots, M_z, W_z>$$

The Public Key is

$$<a_{z,1}, a_{z,2}, \ldots, a_{z,n}>$$

(In fact, the public key is a permutation of the above sequence. This is not a serious problem for the breaking algorithm, but will be ignored in this abstract for the sake of a clearer exposition.)

III. Breaking the Iterated Knapsack

Below we outline the method for breaking the iterated knapsack. Many details are omitted for the sake of clarity.

We will show how to recover $W = W_z$ and $M = M_z$. By iterating the process, the entire secret key can be obtained.

Let $L = L_z$ denote the inverse of W MOD(M). Clearly it is enough to recover L, M.

We know the following system of congruences holds.

[SI] $La_{z,i} \equiv a_{z-1,i} \text{MOD}(M)$ $i = 1, 2, \ldots, b$

(b depends on certain parameters used in constructing the system under attack. For a typical real system $b \simeq 6$.)

Rewriting SI as a system of equalities we get

$$[SII] \quad La_{z,i} - K_i M = a_{z-1,i}$$

for some natural numbers K_i.

The $a_{z,i}$'s are known, since they are part of the key, but all other quantities are unknown. We could solve SII for L and M and the K_i's except that,.

Problem 1

The system is underdetermined and has infinitely many sets of solutions and we need a way of distinguishing the correct one.

Problem 2

The system is non-linear ($K_i M$ terms) and so we have no polynomial time algorithm to solve it anyway.

Curiously Problem 1 will provide a solution to Problem 2.

However, before solving Problem 2, we will simplify SII a bit, by removing some of the unknowns. Note that M is larger than the $a_{z-1,i}$'s by construction. The larger M is with respect to the $a_{z-1,i}$'s the better for the breaking algorithm. Just how much larger is enough will be analyzed in the paper. For now we will assume there is a fixed d (known to us) such that the $a_{z-1,i}$'s $\leq M/d$. This allows us to use SIII below.

$$[SIII] \quad La_{z,i} - K_i M \leq M/d$$

Even with SIII, however, Problems 1 and 2 remain.

Problem 1 says that in addition to the desired solution with L and M, SIII has infinitely many solutions which are undesirable. In fact, solutions are so abundant that we can be sure there will be a solution when

2^c (for a c easily computable from the public key) is substituted for M. That is we know

[SIV] $\quad La_{z,i} - K_i 2^c \leq 2^c/d$

is solvable.

SIV has some good properties and some bad ones.

1. (Good) Since the $a_{z,i}$'s (and c and d) are known, SIV is linear and can be solved (in many cases) using Lovacz-Lenstra.

2. (Good) It can be heuristically argued that the solution for L and the K_i's is unique.

3. (Bad) When solved, the value for L obtained has no useful relationship to the L we want (because the wrong M was used).

4. (Good) The K_i's which are obtained are correct. That is, we can show that they are the same K_i's that we would get if the correct M had been used!

Therefore we can substitute these K_i's into SIII and remove the non-linearity.

Eliminating Problem 1 requires transformation of the problem in a way which will not be described in detail here. The end effect will be that we will be able to augment system SIII with new linear inequalities which state, in essence, that the values $La_{z,i} - K_i M \; (= a_{z-1,i})$ are not only small but "special" in that they are either super-increasing (z = 2) or the iterates of some super increasing sequences. Finally it will be argued that the augmented system has a unique solution -- the desired L and M.

References

[1] Diffie, W., and Hellman, M. E., "New directions in cryptography," <u>IEEE Trans. Information Theory</u>, IT-22, Nov. 1976, pp. 644-654.

[2] Lenstra, A. K., Lenstra, H. W., and Lovacz, L., "Factoring polynomials with rational coefficients," Report 82-05, March 1982, Department of Math., Univ. of Amsterdam.

[3] Merkle, R. C., and Hellman, M. E., "Hiding information and signatures in trapdoor knapsacks," IEEE Trans Information Theory, IT-24, Sept. 1978, pp. 525-530.

[4] Rivest, R., Shamir, A., and Adleman, L., "A method for obtaining digital signatures and public-key cryptosystems," CACM 22-2, Feb. 1978.

[5] Shamir, A., "A polynomial time algorithm for breaking basic Merkle-Hellman cryptosystem," Proc. 23rd Annual Foundations of Computer Science, 1982.

Rump Session: Impromptu Talks by Conference Attendees

LONG KEY VARIANTS OF DES

Thomas A. Berson

Sytek, Inc.
1225 Charleston Road
Mountain View, CA 94043

INTRODUCTION

The Federal Data Encryption Standard (DES) [1] is a block product cipher which converts 64-bit blocks of plaintext into 64-bit blocks of cipher text, or vice-versa, under the control of a 56-bit key. There has in the past been considerable controversy over the adequacy of DES key length [2,3,4]. Easily implemented modifications to the DES key schedule (KS) would allow the use of keys longer than 56 bits.

Of course, an algorithm with a modified KS might not conform with the published standard and should not strictly be called "DES."

LONG KEYS

The operation of DES on a single block of data requires 16 rounds of internal computation. Each round involves 48 bits of key, which are selected from the 56-bit key under control of the KS function. We will follow the notation of [1], where n is the round number, Kn the 48 bits of key used in round n, and KEY is the 56-bit input key.

$$K_n = KS(n, KEY) \quad n=1, \ldots, 16$$

The function KS is not itself a function of either KEY or of the data. Many commercial implementations of DES, especially those using technology where memory locations are inexpensive, precompute all 48x16=768 bits of the Kn in order to reduce the block processing time.

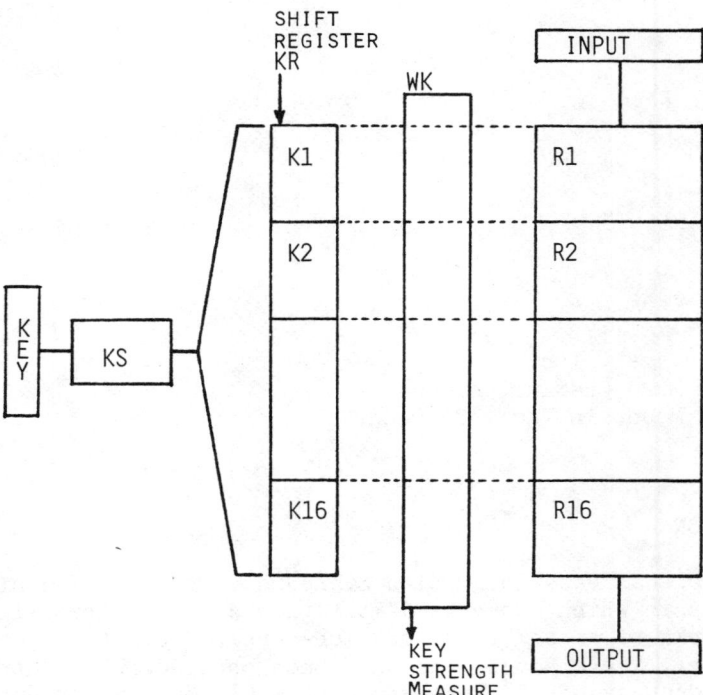

Fig. 1. DES is shown as a pipeline to illustrate the relationship between the Kn and the Rn. Also shown are the proposed placements of KR, a 768-bit key shift register and of WK, a key strength measurement function.

Figure 1 illustrates the relationship between the rounds of DES (R1-R16) and the key (K1-K16). It assumes that all Kn are precomputed and stored in a key register KR. KR is shown as a shift register, although other organizations are also possible (e.g. an array of 16 48-bit words). If the Kn are not stored internally then the same effect can be gained by requiring input of the each Kn at its appropriate round. This may lead to serious timing problems.

No matter how KR is organized, the making of KR (and thereby the Kn) directly accessible allows the user to bypass the standard KS and insert any 768 bits of key.

GENERATING A LONG KEY

How shall the user choose 768 bits of key? One way would be to select a subsequence from a real or pseudo random number generator. Another way would be to devise some new key schedule which generates 768-bit keys from some KEY of different length. A special case of this scheme would be to use the standard KS. This would yield no key length benefit of course, but would provide operation exactly compatible with DES and with other DES devices.

WEAK KEYS

Not all $2^{**}56$ possible KEY values result in equally satisfactory operation of DES. A number of "weak" and "semi-weak" keys have been identified [5]. In general, these keys lack strength because they expose a symmetry or other regularity in KS. Their use results in Kn taking on only one or two distinct values for all n.

The direct provision of a long key avoids the symmetries of KS, but opens fresh opportunities for poverty, repetition, and regularity amongst the Kn. Conceptually, such weak long keys could be guarded against by the definition and implementation of a weak key (WK) function over KR and yielding a measure of weakness. In one scenario, the user would load a long key and compare the output of WK against some predetermined threshold. If the key was weaker than desired it could be altered (e.g. by shifting in additional key bits) until it had sufficient strength.

REFERENCES

[1] Data Encryption Standard, FIPS PUB 46, U.S. Department of Commerce, National Bureau of Standards, (January, 1977).

[2] W. Diffie and M.E. Hellman, "Exhaustive cryptanalysis of the NBS Data Encryption Standard," Computer (10) pp.74-84 (June, 1977).

[3] R. Morris, N.J.A. Sloane, and A. Wyner, "Assessment of the National Bureau of Standards proposed Federal Data Encryption Standard," Cryptologia, (1) pp. 281-291 (July, 1977).

[4] R. Sugarman, "On foiling computer crime," IEEE Spectrum, (16), 7 (July, 1979).

[5] D.W. Davies, "Some regular properties of the DES algorithm," presented at CRYPTO '81, Santa Barbara, CA, (August, 1981).

ON THE SECURITY OF MULTI-PARTY PING-PONG PROTOCOLS

(Abstract)

S. Even and O. Goldreich

Computer Science Department
Technion - Israel Institute of Technology
Haifa, Israel

We define a p-party ping-pong protocol and its security problem, along the lines of Dolev and Yao definition of a two-party ping-pong protocol.

In the case of two parties, it was assumed, with no loss of generality, that there exists a single saboteur in the net and the protocol was defined to be secure iff it was secure against the active interventions of one saboteur. We show that for more than 2 parties this assumption can no longer be made and that for p parties $3(p-2) + 1$ is a lower bound on the number of saboteurs which should be considered for the security problem. On the other hand we establish a $3(p-2) + 2$ upper bound on the number of saboteurs which should be considered. We conclude that for a fixed p, p-parties ping-pong protocols can be tested for security in $O(n^3)$ time and $O(n^2)$ space, when n is the length of the protocol. We show that if p, the number of participants in the protocol, is part of the input then the security problem becomes NP-Hard. Relaxing the definition of a ping-pong protocol so that operators can operate on half words (thus introducing commutativity of the operators) causes the security problem to be undecidable.

INFERRING A SEQUENCE GENERATED BY A LINEAR CONGRUENCE

Joan B. Plumstead

Computer Science Division
University of California
Berkeley, California

ABSTRACT

A pseudo-random number generator is considered cryptographically secure if, even when a cryptanalyst has obtained long segments of the generator's output, he or she is unable to compute any other segment within certain time and space complexity bounds. A pseudo-random number generator which is as cryptographically secure as the Rivest-Shamir-Adleman encryption scheme is presented in [Shamir]. This method for generating pseudo-random numbers is quite slow, though, and it is not known whether any statistical biases might be present in the sequences it generates. Blum and Micali [BlMi] give a pseudo-random bit generator, with arbitrarily small bias, which is cryptographically strong, assuming the problem of index finding is intractable. But their method is also slow. Other cryptographically strong, but slow, pseudo-random bit generators are given in [BBS] and [Yao]. This suggests the question of whether any of the pseudo-random number generators commonly in use are also cryptographically secure. In particular, the linear congruential method, $X_{i+1} = aX_i + b \bmod m$, is very popular and fast. Obviously, this method is not cryptographically secure if the modulus, m, is known. In that case, one could solve for x in the congruence $(X_2 - X_1) \equiv x(X_1 - X_0) \bmod m$. Then the remainder of the sequence could be correctly predicted using $X_{i+1} = x(X_i) + (X_1 - x(X_0)) \bmod m$. In [K1980], Knuth has discussed this problem, assuming m is known and is a power of two, but assuming that only the high order bits of the numbers generated are actually used. We have looked at the problem, assuming the m is unknown and arbitrary, but that the low order bits are also used. We have shown that, under these assumptions, the linear congruential method is cryptographically insecure. A similar result is given in [Reeds], but, among other problems, that result relies on the assumption that factoring is easy.

To reiterate, assume that a fixed linear congruential random number generator, $X_{i+1} = aX_i + b \bmod m$, is given, but the constants a, b, and m are unknown. The problem is to predict, the remainder of the sequence from some of the X_i. First we consider the problem of finding an $\hat{a}$ and a $\hat{b}$ such that $X_{i+1} = \hat{a}X_i + \hat{b} \bmod m$ for all $i \geq 0$, without finding m. The two algorithms we have found to do this use only an initial segment of the sequence. The first algorithm is somewhat easier, but the second is optimal in the sense that it never requires knowledge of more of the X_i than are necessary to completely determine $\hat{a}$. In the worst case, neither algorithm needs an initial segment of length greater than $2 + \lceil \log_2 m \rceil$.

Then, we use a modified version of the second of these algorithms to predict m and the unknown portion of the sequence. Here the problem becomes one of inference. In some cases, it may take a long time to find m, even though many correct predictions for the X_i can be made before m can be uniquely determined. The algorithm we found looks at X_0, X_1, and X_2, and then begins predicting the X_i. We assume that whenever an incorrect prediction occurs, the correct value is revealed before the next prediction is made. In a few cases, X_0, X_1, and X_2 will be sufficient to determine an $\hat{a}$ and a $\hat{b}$ such that $X_{i+1} = \hat{a}X_i + \hat{b} \bmod m$ for all $i \geq 0$. Otherwise, when the first incorrect prediction is made, suitable $\hat{a}$ and $\hat{b}$ are determined from the correct X_i value, as well as an initial guess for m. When further errors are made in predicting the X_i, m is updated so that it is consistent with all previous data. Analysis of the algorithm shows that no more than $2 + \log_2 m$ updates need ever be made.

Finally, we show that, if some of the X_i with $i > j$ are known, they can also be useful in predicting X_j.

The algorithms described above and the proofs of these results can be found in [P1].

REFERENCES

[BBS] Blum, L., Blum, M., and Shub, M., *A Simple Secure Pseudo-Random Number Generator*, Advances in Cryptography: Proceedings of CRYPTO 82, 1982.

[BlMi] Blum, M., and Micali, S., *How to Generate Cryptographically Strong Sequences of Pseudo-Random Bits*, Proc. 23rd IEEE Symp. on Foundations of Computer Science, 1982.

[K1980] Knuth, D.E., *Deciphering a Linear Congruential Encryption*, Technical Report 024800, Stanford University, 1980.

[Pl] Plumstead, J., *Inferring a Sequence Generated by a Linear Congruence*, Proc. 23rd IEEE Symp. on Foundations of Computer Science, 1982.

[Reeds] Reeds, J. *"Cracking" a Random Number Generator*, Cryptologia, Vol. 1, January 1977.

[Shamir] Shamir, A., *On the Generation of Cryptographically Strong Pseudo-Random Sequences*, International Colloquium on Automata, Languages, and Programming, 7th, 1980.

[Yao] Yao, A., *Theory and Applications of Trapdoor Functions*, Proc. 23rd IEEE Symp. on Foundations of Computer Science, 1982.

KEY RECONSTRUCTION

(Abstract)

Michael Merritt
Georgia Institute of Technology

The key reconstruction problem is an application of (n,t)-threshold schemes. Such schemes permit an arbitrary piece of information to be 'broken' into n different 'fragments,' any t of which may be used to completely reconstruct the original information. No smaller number of fragments provides any information concerning the original, aside from its size in bits. These n fragments may then be distributed to different individuals, any t of whom can later cooperate in the reconstruction of the original information. For our purposes, the original information might be a private key in a public key cryptosystem.

Should it become necessary to reconstruct such a key, following the creation and distribution of its fragments, it could be important that only those individuals who where trusted with a fragment receive a copy of the reconstructed key. Under ordinary circumstances, the recipients of the key fragments would be aware of each others identity, or would be given some authenticating information along with their particular key fragment. This information could then be used to ensure that the key be reconstructed without being divulged to anyone who did not originally hold a key fragment. Indeed, error-correcting codes may be used to place this authenticating information with the key fragments themselves. The problem becomes much more difficult, however, if the key fragments can only be authenticated following reconstruction of the key. This situation could arise if the only authenticating information available is a public message signed with the private key, and reliably broadcast to the system users. A key fragment could then be authenticated only by reconstructing the private key with t-1 other valid fragments. Failing to construct the key with t candidate fragments would indicate that at least one of them was invalid -- but not which one.

The key reconstruction problem is to design a protocol for a group of individuals who each claim to hold a fragment of the original key, with the following restrictions; if there are at least t valid fragments, everyone holding a valid fragment must eventually obtain the reconstructed key, and the remaining individuals, no matter what their behavior during the execution of the protocol, must together obtain at most t-1 valid fragments, and thus be unable to reconstruct the key, even acting as a group.

Both upper and lower bounds have been obtained for protocols solving the key reconstruction problem. These bounds are sensitive to three parameters, (n,b,t), where n is the total number of participants claiming to hold valid keys, b is the number who actually holding invalid fragments, and t is the threshold (the number of valid fragments needed to reconstruct the key). These results are summarized in the table below. The algorithms and proofs will appear in the author's Ph.D. thesis.

Restrictions on (n,b,t)	Upper Bounds	Lower Bounds
t even	b+2	b+1
t odd	b+1	b
b=1	b+1	b+1
(7,4,3)	b	b
(6,2,4)	b+2	b+2 (conjectured)

NONDETERMINISTIC CRYPTOGRAPHY

Carl R. Nicolai

Cryptext Corporation
Box 425 Northgate Sta.
Seattle Wa. 98125

Abstract— By employing a special type of stutter and executing it at truly random places in a key stream, the complexity of the stream can be vastly enhanced.

Since the evolution of code breaking by frequency analysis, it has been obvious that the injection of various truly random elements into a text rendered codebreaking more difficult. Early codes made use of nulls to accomplish this purpose. I am presenting what I believe to be new ways of randomness injection into key stream cypher systems. Since linear generators have been widely studied I will use them to illustrate these methods.

Feedback shift register generators based on primitive polynomials have a known break length. Figure #1 shows such a generator and its output. 2N bits defines the generator and N bits can completely key it.

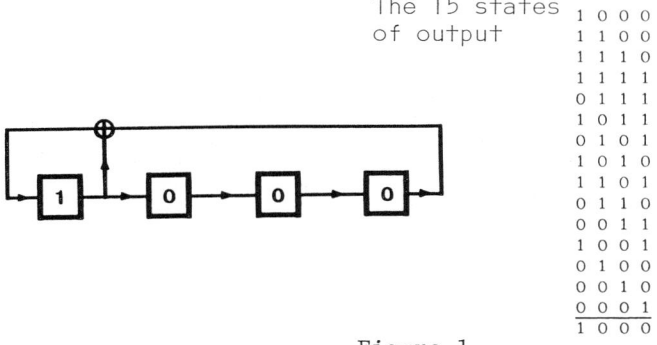

Figure 1

If N bits are known (or can be guessed at) the subsequent output of the generator can be predicted.

Several sub-generators can be EXORed together (MOD 2 added) to extend their break length, and if the periods of these sub-generators are relatively prime a single stream will be produced. The resultant generator, while not maximal will still be very long. The break length then is the sum of the number of bits contained in the sub-generators.

Figure #2 shows several relatively prime maximal period generators connected together. The resultant stream can be viewed as one long line of output bits.

There are several ways of jumping from one part of the stream to another. One way of moving a known large distance in the stream is to tick (advance the generator 1 bit) one of the sub-generators and not the others. Modifying the contents of the sub-generators is another. I will use the term Code Branch to indicate a special type of stream hopping (a type of stutter). The Code Branch will have the following properties.

    1. It will move the output to a part of the stream a very long distance from the previous state.
    2. Exactly how far the output is moved will depend on the previous state.
    3. The Branch will destroy information so that the pre Branch state cannot be determined. (a one way valve)

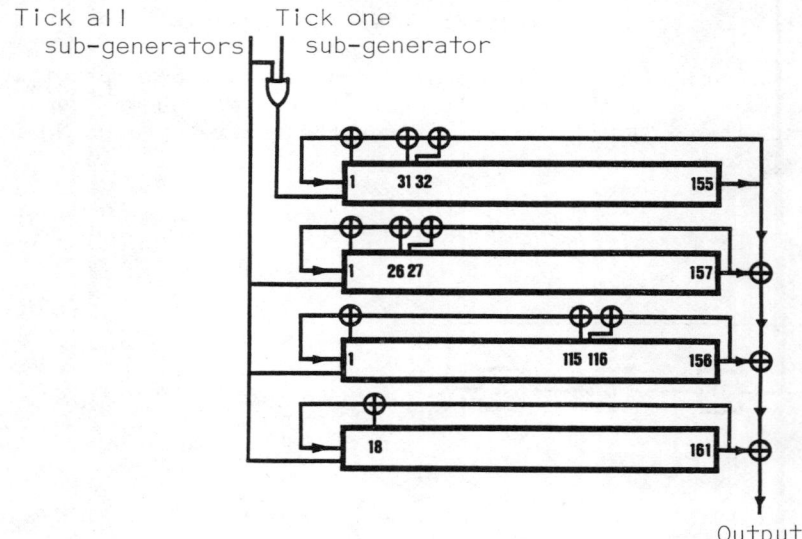

Figure 2

Figure 3 illustrates a stream which is branched at state x, x+1, and x+2 locations. The distances between the post branched states should be large enough to prevent overlapping.

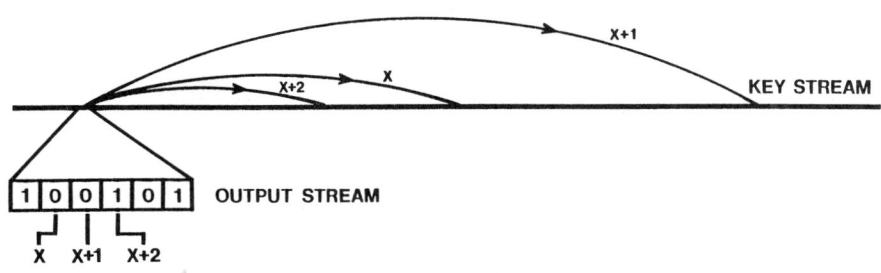

Figure 3

The generator in figure 2 requires 629 bits to key it. If we let the generator tick 529 times 100 bits would still be required to key (or break) the generator. In other words the generator could be in any one of approximately 2 to the 100 states and still have produced the same 529 bit output stream.

Now if we execute a code branch between say 400 and 529 states the resultant state of the generator could be in any one of 129 locations, and have an uncertainty of 100 to 229 bits.

It can be easily shown that the outputs of generators that have been randomly branched can contain sequences of digits that could not have been produced by the original generator, but only by one that is much longer. Thus it may be argued that random branching of generators produces generators of indeterminate length. Clearly such generators are harder to cryptoanalyze then if branched by deterministic methods, which in general only complexify the output to the fixed level of the branching algorithm.

Randomly jumping between different generators can also produce streams of output with a break length longer than the combined length of the generators.

To use a randomly branched stream in a crypto system bandwidth expansion is required. In the first example we will make the transmitted character set larger than the valid character set. By expanding the 7 bit ASCII character set to 8 bits for example we make the probability of decoding an incorrect character one half. Thus if we are decoding a message that has been encrypted with a randomly branched stream and we detect an error character we know that somewhere before this character a code branch must have been inserted.

It is a simple matter to restart the decoding again and insert code branches further and further back until no more error characters occur. Using a trailing block of characters insures that the complete message will be correctly decoded. By using a leading block of truly random characters, and insuring that the code branch always executes in a different place in this first block, any number of cyphertexts can decode to the same message.

A more bandwidth efficient method is to have the message and transmitted characters the same except for reserving one of the message characters as an indicator of when a code branch is going to occur. When this character is decoded the decoder inserts a branch prior to decoding the rest of the message.

One of the things that can be done with code branching is to adjust the probability of either the key stream or the transmitted characters. If the encoder looks at the stream and branches at a random loacation but before a certain byte (say 8 zeros) occurs then the statistics of the stream have been dewhitened in a manner that insures that all encrypted bytes will have at least one bit of difference from the plaintext. This process will involve multiple encryption passes.

If code branches occur at planned intervals then two messages may be interleaved in the cyphertext. One message would be the decrypted cyphertext and the other would be decoded by where the code branches occurred.

A SHORT REPORT ON THE RSA CHIP

Ronald L. Rivest

MIT Laboratory for Computer Science
Cambridge, Mass. 02139

The nMOS "RSA chip" described in our article [1] was initially fabricated by Hewlett-Packard. Testing revealed that while the control portion of the chip worked correctly, the arithmetic section suffered from transient errors and was usually too unreliable to complete a full encryption. We tested a number of chips and found the same problem, enough to convonce us that the cause was probably a design error and not a fabrication problem.

We have recently refabricated the chip, through the kind auspices of Xerox, with an improved power distribution network. While we had good reasons to believe this would help, it was disappointing to find out that this modification was insufficient to prevent the transient errors.

We are hopeful that we can identify and correct the bug in the near future.

The reader may be interested to know that a CMOS RSA chip of a rather different architecture is likely to be available soon as a commercial product; interested parties may contact the author for further information.

References

[1] Rivest, R. L., "A Description of a Single-Chip Implementation of the RSA Cipher," *LAMBDA Magazine* 1, 3(Fourth Quarter 1980), 14-18.

Author Index

Adleman, Leonard M., 259, 303
Akl, Selim G., 237
Avis, G. M., 139
Bennett, Charles, 267
Berson, Thomas A., 251, 311
Blakley, G. R., 39
Blom, Rolf, 231
Blum, Lenore, 61
Blum, Manuel, 61
Brassard, Gilles, 79, 267
Breidbart, Seth, 267
Brickell, Ernest F., 15, 51, 289
Chaum, David, 199
Davies, Donald W., 89, 97
Davis, J. A., 289
Dolev, Danny, 167, 177
Even, Shimon, 177, 205, 315
Goldreich, Oded, 205, 315
Goldwasser, Shafi, 211
Hellman, Martin E., 3, 129
Janardan, Ravi, 21
Jueneman, Robert R., 99
Karp, Richard M., 177
Lakshmanan, K. B., 21

Lempel, A., 205
Longpré, Luc, 187
Merritt, Michael, 321
Micali, Silvio, 211
Moore, J. H., 15
Mueller-Schloer, Christian, 219
Nickolai, Carl, 323
Parkin, G. I. P., 97
Plumstead, Joan B., 317
Reyneri, Justin M., 3, 129
Rivest, Ronald L., 145, 327
Shamir, Adi, 279
Sherman, Alan T., 145
Shub, Michael, 61
Simmons, Gustavus J., 289
Swanson, Laif, 39
Tavares, S. E., 139
Taylor, Peter D., 237
Wagner, Neal R., 219
Weisner, Steve, 267
Wigderson, Avi, 167
Winternitz, Robert S., 133
Yao, Andy, 211

Subject Index

This index associates each entry with the initial page number of every paper that relates to the entry.

Access control, 251
Authentication tags, 79
Blind signatures, 199
Byzantine Agreement, 167
Crusader Agreement, 167
Cryptanalysis, 279, 289
Data Encryption Standard (DES)
    cycle length, 97, 99
    ouput feedback mode, 89, 99, 311
    statistical regularities, 89, 129
    weak keys, 89, 99, 311
Digital signatures, 21, 79, 187, 199, 211
Discrete logarithms, 3, 15
Electronic funds transactions, 187
Electronic notary public, 259
Finite fields, 3, 15
Graham-Shamir cryptosystems, 303
Information theory, 39
Key distribution systems
    convential, 231
    public-key, 3, 15
Keystream ciphers, 133
Knapsack systems, 279, 289, 303
Matrix cover problem, 21
Merkle-Hellman cryptosystems, 279, 289, 303
Multilevel security, 237
Multiplication algorithms, 51
NP-completeness, 21
Nuclear test ban treaty, 259
Oblivious transfer, 205

One-time pad, 39
Output feedback mode, 97, 99
Personal data cards, 219
Polarized light, 267
Protocols
    multi-party, 167, 315
    network, 251
    ping-pong, 177, 315
    randomized, 205
Pseudo-random number generators, 61, 317
Public-key cryptography
    cryptanalysis of knapsack systems, 279, 289, 303
    cryptosystem based on matrix cover problem, 21
    key distribution systems, 3, 15
    use in creating unforgeable tokens, 267
    use in implementing notary publics, 259
    use in signing checks, 187
Quantum mechanics, 267
Random functions, 89, 129
Random graphs, 129
Randomized encryption, 139, 145, 323
Randomized protocols, 205
Redundant number systems, 51
RSA chip, 327
Threshold schemes, 39, 321
Universal Hashing, 79
Untraceable payments, 199
Verify only memory (VOM), 267
Very large scale integration (VLSI), 327